706

SIXTH EDITION
EXERCISES IN ENVIRONMENTAL SCIENCE

MICHAEL SLATTERY
KRISTI ARGENBRIGHT
KRISTEN DEBONE
MICHAELA DONOHOO

Kendall Hunt
publishing company

Cover images © Shutterstock, Inc.

All graphs/figures are supplied by author.

Kendall Hunt
publishing company

www.kendallhunt.com
Send all inquiries to:
4050 Westmark Drive
Dubuque, IA 52004-1840

Contents

Lab 1

Your Ecological Footprint

Introduction

Welcome to Contemporary Issues in Environmental Science. Throughout this semester, you will learn about many of today's environmental issues and discuss the "what?," "how?," "why?," and the "so what?" of each issue we cover. But first, we need to understand three main concepts before we begin to discuss today's environmental problems and their potential solutions.

First, many of these issues are interconnected. Most environmental challenges facing the global community are not isolated within themselves. Conversely, many of these challenges are, in fact, integrated across many scales. For example, we cannot discuss the problem of our rapidly increasing global population without discussing the growing demand for food, water, and energy.

Next, finding solutions to environmental issues is necessarily an interdisciplinary practice. It requires input from the scientific (natural and social science) community, political leaders, and the global community, all of whom contribute to and are affected by these issues.

Finally, solutions to environmental issues must be sustainable. A sustainable method utilizes a resource without depleting or permanently damaging it. As stewards of the earth, living sustainably means that we should, at times conserve, preserve, and allow environmental regeneration of resources, so that future generations can become equal stewards of the resources vital to life.

All the scientific research in the world is just a pile of data collecting dust unless people are willing to implement those findings into policy solutions and into their daily lives. A big hurdle for generating sustainable solutions is getting citizens to consider environmental issues as personal issues. Due to the sheer size of many of the issues, individuals often feel powerless to affect any positive change.

Through the exercises in this lab manual, we hope to not only show you the science behind relevant environmental issues, but also to help you develop a sense of responsible environmental stewardship through open discussion of how even your daily personal choices can make a difference in the rising issues.

YOUR ECOLOGICAL FOOTPRINT

One of the most puzzling environmental questions comes when determining how many people Earth's resources can support. This brings us to the concept of *carrying capacity,* or the maximum number of individuals in a population that an ecosystem can support without being adversely impacted (we will discuss the concept of carrying capacity further in Lab 2).

As a PhD candidate, Dr. Mathis Wackernagel developed the concept for the **ecological footprint** in 1992 alongside William Rees, his professor at the University of British Columbia. The goal of the ecological footprint was to measure the impact humans were having on the planet at the individual scale. The result was a tool referred to as the ecological footprint quiz, which allows you to see how your daily choices impact the planet. The quiz estimates how many acres of **ecologically productive land area** are required to support your lifestyle, based upon what you eat, where you live, and how much you shop and travel. The results are expressed in global acres (or global hectares in metric measurement). Each acre corresponds to one acre of ecologically productive space with world-average productivity. Today, the biosphere has 26.7 billion acres (or 10.8 billion hectares) of ecologically productive space corresponding to less than one-quarter of the planet's surface. These 26.7 billion acres (10.8 billion hectares) include the following:

- 5.7 billion acres (2.3 billion hectares) of productive ocean
- 21 billion acres (8.5 billion hectares) of productive land

When the total amount of ecologically productive land area available today is divided by the human population, there are about *4.5 acres (1.89 hectares)* available for each person.

Ecological footprints can also be measured on a larger scale to document a given population's consumption and waste production, and are expressed in global acres of ecologically productive space necessary to maintain these services. The **biocapacity** of a given area, such as a single country, is the capacity of an ecologically productive area to generate an ongoing supply of renewable resources and to absorb its spillover wastes.

By aggregating data on the consumption of various resources and on the ability of the earth to provide them, ecological footprint accounts (EFA) provide a means to compare various components of consumption and ultimately serve as an indicator of sustainability or, in the case of deficits, of unsustainability.

BOX 1.1   Industrialized Economies, Industrializing Economies, and Economies in Transition

Throughout the semester, you will be asked to think about the economic and environmental conditions in different countries around the world. It is important to take into account the true state of the economies in these countries before trying to compare them with one another.

Until recently, it was easy to think about the world's nations by separating them into two different categories: industrialized nations and industrializing nations. Industrialized nations (or "developed countries") are countries with a high level of economic development and security, and are generally characterized by the use of advanced technologies and higher standards of living. Industrializing nations (or "developing countries") are still in the process of developing an industrialized economy and generally feature a large agrarian (farming) economic sector and lower standards of living. Industrialized nations include the United States and many countries in Western Europe, while industrializing nations include countries such as Pakistan, Kenya, Costa Rica, and Chile.

However, after the end of the Cold War and the dawning of the Age of Globalization, there has been a closing of the gap between industrialized and industrializing nations. The result has been a greater number of countries that lie in the middle of the spectrum. Many of these countries are continuously undergoing modification of their formerly state-controlled (or government-owned) economic structure to better resemble a free market structure. These countries are often referred to as "Economies in Transition" and include China and many of the countries in Eastern Europe. Other countries, such as India and Brazil, have rapidly developed an export-based economy and increased the living standards for certain portions of their population in the process, although poverty is still widespread in other parts of these countries. Many "middle of the spectrum" countries are still often referred to as developing countries. However, for the sake of our analysis it is important to remember that these countries are increasingly distinguished from nations with less-developed economies.

Individuals that live in industrialized countries will generally have a greater ecological footprint than those who live in less economically developed countries. Those who live in cities will also have a larger footprint than those who live in rural areas. While much of what the ecological footprint calculates is based upon your individual choices, there are some aspects of your footprint that are out of your immediate control. Access to public transport, renewable energy, and energy-efficient housing may not be available to you, they are nevertheless calculated into your footprint.

According to Dr. Wackernagel's "Global Footprint Network," we are in **ecological overshoot**.[1] That means we are living beyond the means of our planet to support us. Western Europe and North America have the highest per capita footprints of any region on Earth. *At 23.6 acres (9.57 hectares) per person, the United States has the largest per capita ecological footprint on the planet.* If, for example, every person in the world were to live like the average American, it would take 5 Planet Earths to support the human population!

Humanity's collective ecological footprint breached the sustainability mark for the first time in the early 1970s and has remained unsustainable ever since. By 2000, the ecological deficit reached nearly 1 acre per person, or 9 million square miles. Moreover, the ecological footprint methodology does not capture all of humanity's impacts on the environment. Toxic pollutants and species extinction, for example, are not incorporated into the footprint model. This is important because the "value" of nature extends far beyond the goods and services that humans take from it. The "footprinting" methodology does, however, offer one of the most comprehensive assessment techniques that can help inform, educate, and point the way toward a more sustainable path.

For now, don't worry about the details behind the ecological footprint quiz (if you're interested, these may be found at *(https://www.footprintnetwork.org/resources/).* Let's go ahead and complete it—we will then discuss the ramifications of your final answer later.

YOUR WATER FOOTPRINT

While the ecological footprint measures your use of an array of resources, your water footprint measures your **direct** and indirect use of fresh water resources. Dr. Arjen Hoekstra developed this footprint in 2002

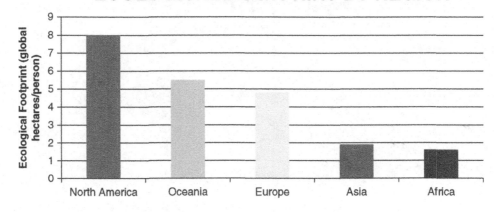

Figure 1.1. *Footprints vary widely by region.*
Source: Data from 2007. Global Footprint Network, Ecological Footprint Atlas, 2010.[2]

---

[1] The Global Footprint Network. 2012. *Global Footprint Network 2012 Annual Report.*

[2] The Global Footprint Network. 2010. *Ecological Footprint Atlas 2010.*

for the United Nations Educational, Scientific, and Cultural Organization (UNESCO). Its purpose was to serve as:

> ... an indicator of water use that looks at both direct and indirect water use of a consumer or producer. The water footprint of an individual, community or business is defined as the total volume of freshwater that is used to produce the goods and services consumed by the individual or community or produced by the business.[3]

Water is often thought to be part of the **commons**, or a resource that is used in common by many people (water, air, grazing land). Access to clean water is vital to human survival and is also a part of every aspect of our day-to-day life. Beyond the water we drink, bathe in, and use for cooking, *water* plays a role in the manufacturing of our homes, clothing, consumer products, and even in the electricity we use. Biologically, humans really only need 2–3 liters per day (< 1 gallon) yet daily American water use is estimated to be 2,000 gallons per day—more than twice the global average!

Water cycles through various reservoirs in the hydrologic cycle, making it a renewable resource. However, water is also a *finite* resource, meaning there is only a certain amount of water on Earth. While most of the planet is covered by water, less than 1% of it is in a form and location that is available to us as humans. Since this 1% is finite, we must strive to conserve the water we have and protect it from pollution and overuse.

Much of water we use is not for direct household activities or consumption. The indirect use of water to produce goods or services is called **virtual water**. For example, it takes more than 2,000 gallons of water to produce just one pound of beef. This water is used not only to sustain the animal and service the farmhouse, but also to grow and transport the grains used to feed the cow. Since the average American eats about 62.5 pounds of beef every year, this means that the average American also consumes about 125,000 gallons of water every year just by eating beef! This virtual water is included in your water footprint.

Often, the virtual water you use does not originate in your own country. The water that is used in other countries to create the products and services you import is included in your **external water footprint**. Conversely, the water you use that is sourced in your own country is included in your **internal water footprint**.

---

[3] Hoekstra, A. Y. 2008. "Water Footprint: Introduction." The Water Footprint Network. http://www.waterfootprint.org/?page=files/home

Name:

LAB 1 WRITE-UP: ECOLOGICAL & WATER FOOTPRINTS

Exercise 1: Ecological Footprint

1. Visit the following link: *http://www.footprintnetwork.org/* to take the quiz. If you live in a residence hall, complete the quiz as if you were living at home with your family.
   Record your results below. (0.25 point)

   Many activities impact our Footprint. If everyone lived like you, we would need _____ Planet earths to provide enough resources.

   To support your lifestyle, it takes _____ global hectares of the Earth's productive area.

2. In one or two paragraphs (**at least four sentences**), answer the following questions: Where is the bulk of your footprint from? Discuss the aspect of your personal ecological footprint you found most intriguing or interesting. (**1 point**)

3. Discuss potential economic, political, or social problems individual cities/towns might face if everyone on Earth had an ecological footprint similar to your own. (**1 point**)

4. After receiving your final results, see how you can reduce your footprint by retaking the quiz by selecting "Retake the Quiz" button in the bottom right-hand corner. Using the "Add Details to Improve Accuracy" see if you can figure out how to actually reduce the size of your footprint quite significantly. Once you are happy with the reduced footprint, you can select "Explore Solutions" to learn how communities around the world are shrinking their personal footprints. **Then write a paragraph or two (at least four sentences) that answers these questions: What are some plausible steps you could take to reduce various aspects of your footprint? How much would these steps reduce your footprint? (1 point)**

5. Go to the following link: *http://data.footprintnetwork.org* to look at footprints from other countries. Select a developing world country (i.e., a country that is not industrialized and a developed country) **other than the United States. Record your countries' names, footprints, and biocapacities below. (0.25 point)**

Developing country:
_____
Ecological footprint: _____ global hectares per person
Biocapacity: _____ global hectares per person
Ecological Footprint: _____ number of Earths

Developed country:
_____
Ecological footprint: _____ global hectares per person
Biocapacity: _____ global hectares per person
Ecological Footprint: _____ number of Earths

6. **Write a paragraph (<u>at least four sentences</u>) comparing the ecological footprints of the two countries that you chose. Make sure to answer the following questions in your answer: How do these two countries compare? Are they similar? What do you know about the countries that could explain their graphs? How do their ecological footprint values compare to the global average of 1.89 global hectares?** For extra information, try visiting the CIA Fact book to learn more about your countries (https://www.cia.gov/library/publications/resources/the-world-factbook). **(2 points)**

Exercise 2: Water Footprint

7a. Visit http://www.watercalculator.org/ and take the water footprint quiz. Again, if you live in a dormitory, please complete the quiz as if you were living at your permanent residence. (Note: Do not hit the "back" button on the browser during the quiz. Let your instructor know if you continue to have trouble completing the quiz after several attempts. **Record your personal water footprint results in the table below.**

7b. Next, compare your water footprint results to the US Average shown in the table below. Are you above or below the US Average? **Mark + if you are above the national average, or – if you are below the national average for each of the areas listed in the table below. (7a and 7b: 0.25 point)**

Indoor Water	U.S. Average (gallons per person per day)	YOU (gallons per day)	Are you above or below the U.S. Average? +/−
Shower	11		
Bathtub	2		
bathroom sink	3		
Toilet	14		
kitchen sink	7		
Dishes	1		
Laundry	10		
Greywater	−25		
Outdoor Water			
lawn & garden	72		
rain barrel	−2		
swimming pool	23		
car washing	1		
Virtual Water			
Driving	5		
Electricity	30		
shopping habits	583		
Paper	−3		
plastic	−1		
bottles & cans	−8		
Fabrics	−1		
Diet	1,063		
pet food	48		
Total	1,802		

8. Discuss potential economic, political, or social problems individual cities/towns might face if everyone on Earth, so not just USA, had a water footprint similar to your own. (**1 point**)

9. In one or two paragraphs (at least four sentences), answer the following questions: Discuss the aspect of your personal water footprint you found most intriguing or interesting. What can you do to lower your water footprint? (Click "view tip" and explore http://environment.nationalgeographic.com/environment/freshwater/water-conservation-tips/ to help get ideas.) (**1 point**)

10. Go to https://www.waterfootprintassessmenttool.org/national-explorer/ and find the water footprints for the two countries you looked at in Exercise 1. If no data is listed for your selected country, selected another country. Note: The website gives per capita water footprints in L/day. You will need to use the conversion below to convert information into gallons per day. Record your countries' footprints and the part of footprint falling outside of the countries below. (**0.25 point**)

Developing country:
Average water footprint per capita: _____ L/day
Conversion: _____ L/day × 0.26 = _____ gal/day
Percent of water footprint falling outside of country (external): _____

Developed country:
Average water footprint per capita: _____ L/day
Conversion: _____ L/day × 0.26 = _____ gal/day
Percent of water footprint falling outside of country (external): _____

11. Compare the two countries to each other and to the global average (global average water footprint = 365,640 gal/yr per capita). In one or two paragraphs (at least four sentences), answer the following questions: Why do you think they have the water footprint they have? What do you know about these countries that might explain their internal and external footprints? (Hint: "part of footprint falling outside of the country" is the external water footprint.) Take into account economic status, import and export patterns, and geography. (**2 points**)

The Global Footprint Network. 2012. *Global Footprint Network 2012 Annual Report.*

The Global Footprint Network. 2010. *Ecological Footprint Atlas 2010.*

Hoekstra, A. Y. 2008. "Water Footprint: Introduction." The Water Footprint Network. http://www.waterfootprint.org/?page=files/home

Lab II

Population

Introduction

Before the Industrial Revolution, the global human population was about 300 million people. By the end of the Industrial Revolution, the human population had increased to 1 billion people, and by October 2011 the human population had crossed over the 7 billion mark. The United Nations predicts that the global population will exceed 9 billion by 2050.

Almost every country in the world now faces problems of population control and the consequences of overpopulation. An overabundance of people results in fewer resources available to all. In fact, some governments have been so concerned with overpopulation in their country that they have declared a state of emergency. On the other hand, some countries (such as Germany) are now faced with a contracting, aging population. Both issues—a rapidly expanding or a contracting population—can be detrimental to a nation's economy and political stability. Furthermore, many foreign policy and security analysts recognize population growth as a threat to national security.

In this lab, we will explore principles behind population dynamics and, more specifically, the human population.

POPULATION DYNAMICS

In order to have an informed discussion on human population, we must first examine some basic concepts that can be used to describe populations from the smallest bacteria to humans.

Growth Curves

To understand trends in population growth, we need to understand the basic terminology of population dynamics. **Demography** is the study of human population and the changes that take place among populations around the world. Demographers are those who record and graph the changing trends of populations over time. The simplest way of recording this data is through *growth curves*, which represent population sizes over time.

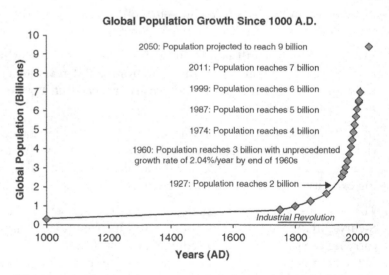

Figure 2.1. *Trends in human population growth over the last millennia.*

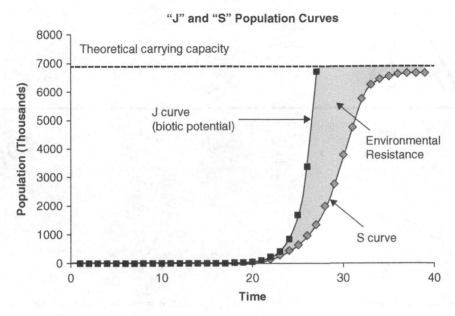

Figure 2.2. *The two main growth curve trajectories: J-curve (also called the exponential curve) and S-curve (also called the logistic curve).*

There are two basic types of growth curves. The first is a **J-curve**, sometimes referred to as a population's *biotic potential*, because it represents the trajectory of a population's growth if there were no constraints to curb the growth rate. Demographers often refer to the J-curve as an exponential growth curve, since it resembles a mathematical exponential curve (Figure 2.2).

The second growth curve is an **S-curve**, which is aptly named for the shape that it resembles (Figure 2.2). The S-curve assumes that there are constraints limiting the population from taking an exponential J-curve trajectory, especially as the population nears its carrying capacity. **Carrying capacity** refers to the number of individuals within a population that can be supported by the ecosystem. Most often, the forces that bend the top portion of the S-curve and result in the population leveling off near carrying capacity arise from *environmental resistance*.

ENVIRONMENTAL RESISTANCE

All populations require resources to support them. As humans are arguably the most successful species on the planet (in terms of expansion and survival), it is often hard for us to remember our dependence on natural resources.

Every species on Earth is influenced by the ecosystem in which they live. Pressure exerted on a population by the ecosystem is known as **environmental resistance**. There are two main types of environmental resistance—*density-dependent* and *density-independent*. Both of these have the ability to influence a population and its growth.

In a theoretical environment void of environmental resistance, a population would increase exponentially (refer to the J-curve discussed earlier). However, we know that in a more practical scenario each ecosystem has a carrying capacity determined by the availability of natural resources. When the population surpasses carrying capacity, they will be met with environmental resistance and growth should begin to level off.

The relationship between exponential population growth and the limit of carrying capacity is expressed as an S-curve. In the S-curve, population growth increases exponentially, then levels off and steadily approaches the carrying capacity without exceeding that limit. If we take a closer look to the steady approach of the S-curve to the carrying capacity, we would see events of population **overshoots**, in which the population exceeds the carrying capacity, and **diebacks**, in which environmental resistance stress factors return the population to below the carrying capacity (Figure 2.3).

In the following sections of this lab, we will discuss the two types of environmental resistance and population growth rates. We will repeatedly use the classic example of a community of foxes and rabbits to provide clarity on these concepts. In this example, think of an ecosystem populated by rabbits, which eat plants. In turn, the foxes in the community (top of the food chain) eat the rabbits.

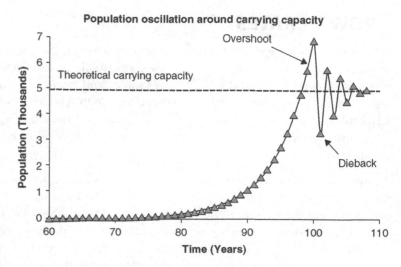

Figure 2.3. *Population overshoots and diebacks, as a result of environmental resistance when the population size exceeds the carrying capacity.*

Density-Independent Environmental Resistance

Density-independent resistance impacts population size regardless of how close the population is to reaching carrying capacity. This form of environmental resistance tends to arise from major weather events that would occur and impact the population regardless of size, such as a hurricane, tornado, or volcanic eruption. Density-independent environmental resistance often leads to a rapid die-off of a population and a fairly rapid rebound response by the population once favorable conditions return.

To refer back to our example, imagine a drought struck the ecosystem where the rabbits and foxes live. Much of the grass would dry out and die, resulting in less food for the rabbits and ultimately less food for the foxes (as the rabbits too begin to die from starvation). This limited amount of food exerts environmental resistance on each of the populations. However, this drought would have occurred whether the population density was far above or far below the carrying capacity of the environment. The drought is an example of density-independent environmental resistance.

Density-Dependent Environmental Resistance

Every population requires natural resources to support them, including food, water, living space, and clean air to breathe. As populations expand and place increasing demands on resources, the environment begins to exert pressure on the growing number of organisms. The level of **density-dependent environmental resistance** is a function of how close the population is to reaching or surpassing the environment's carrying capacity. When a population approaches carrying capacity, the environmental resistance intensifies and suppresses the population's growth rate in order to return the population to sustainable numbers (S-curve trajectory). Ultimately, this means that the greater the number of individuals in a population, the greater the environmental pressure exerted upon that population.

Remember that in our example food chain the foxes eat the rabbits and the rabbits eat the plants. What happens if the rabbits eat all the plants? Limited food availability will exert environmental pressure on the rabbit population and some rabbits will die off. Fewer rabbits mean less food for foxes. Therefore, as plant abundance exerts environmental resistance on the rabbit population, rabbit abundance exerts environmental resistance on the fox population. This type of interaction is known as a bottom-up control. **Bottom-up controls** affect a population based upon the availability of food and other required resources.

Conversely, what if the foxes all went away? The rabbit population would grow out of control and overgraze on the plants until they ran out of food. This second type of interaction is known as a top-down control. **Top-down controls** result from populations at the top of the food chain exerting pressure on populations below it on the food chain, typically through predation. In other words, the fox population is responsible for controlling the rabbit population.

GROWTH RATES

All populations have optimal environments in which they can thrive and grow. When a population is in a resource-rich environment, it will grow exponentially or experience a population boom. As that population expands, however, it will drain those resources and begin to create a less optimal environment with fewer resources. A population's response to this decrease in resources is typically limited to one of three responses: adaptation, migration, or death. While environmental resistance and a population's response may be difficult to quantify, *growth rates* can give an indication of the level of environmental resistance being exerted on a population.

A **population growth rate** is simply the rate at which the number of individuals in a population increases from the initial population size in a set time period. At the most basic level, a growth rate is calculated based on the number of individuals entering a population and the number of individuals leaving the population (Eq. 2.1). It is the first and most imperative method of quantifying a population's growth or demise. Growth rates can assist in determining the level of environmental resistance exerted onto a population. Expressed as a percent, growth rates are perhaps the most universal descriptor of population dynamics. When the growth rate is positive, the population is increasing; if the growth rate is negative, then it is decreasing. We will explore three different types of formulas used to calculate growth rates.

Because it is often impossible to get an exact census for a population (a physical count every individual), population scientists instead utilize **representative samples**, or subsets of the population. When calculating growth rates, it is important to remember that the sample size (denominator) must be consistent.

General Growth Rate

The generalized growth rate formula is a simplified formula used to give us a basic understanding of a population's growth. It solely focuses on the number of individuals coming into the population and the number of individuals leaving a population.

$$growth\ rate\ (r) = \left(\frac{(incoming - outgoing)}{sample\ size} \right) \times 100$$

(2.1)

The next two formulas are variations on the general growth rate formula.

Rate of Natural Increase

Finding the **Rate of Natural Increase (RNI)** is a way to determine the growth rate of an *isolated* population— one that is not influenced by immigration or emigration. To calculate RNI, simply subtract the number of deaths (outgoing) in a population from the number of births (incoming) before dividing the difference by the sample size (Eq. 2.2).

$$RNI\ (r) = \left(\frac{(births - deaths)}{sample\ size} \right) \times 100$$

(2.2)

It is important to note that RNI requires that births and deaths be expressed with the same sample size. For example, you couldn't calculate RNI if your birth rate was 20 per 1,000 individuals and your death rate was 10 per 2,000 individuals. You would have to use identical sample sizes and multiply the births and/or deaths accordingly to express the correct proportion.

Population Growth Rate

The **Population Growth Rate (PGR)** considers immigration and emigration in addition to births and deaths when quantifying growth rates (Eq. 2.3). In contrast, RNI only considers births and deaths in a population, so RNI would not be appropriate when trying to describe a population that gains or loses a portion of its population from **immigration** (individuals migrating into the study population) or **emigration** (individuals migrating out of the study population). This growth rate is ideal for populations that are not geographically isolated.

$$PGR = \left(\frac{[(births - deaths) + (immigration - emigration)]}{sample\ size} \right) \times 100 \qquad (2.3)$$

Again, the sample size must be the same or must be converted so that each metric uses the same sample size.

DOUBLING TIME

In addition to knowing growth rates, it would also be helpful to know when the population would double or halve in size. Doubling time can be especially useful when calculating human populations because a doubled population requires double the resources to meet the larger population's needs.

Doubling time (T_d) expresses the amount of time it would take for a given population to increase by a factor of 2. Doubling time employs the *Rule of 70*, which is commonly used in finance to estimate when an investment that compounds annually would double. To calculate when a population would double in size you would divide 70 by the growth rate (either RNI or PGR) (Eq. 2.4).

$$T_d = \frac{70}{growth\ rate} \qquad (2.4)$$

Doubling time refers to a positive growth rate. However, if the growth rate is negative, the same equation would yield the population's **halving time**, or the time it would take for the population to be decreased by half.

While growth rates are expressed as a percent, to calculate doubling time you treat the percent as a number. That is, if your growth rate is 2.0% you would divide 70 by 2.0 rather than by 0.02.

$$T_d = \frac{70}{2.0} = 35\ years$$

Interestingly, while the change in growth rate may seem small, say from 2.0% to 1.5% to 1.0%, it can have a large impact on the doubling time. Consider:

Growth rate	Doubling Time
2.0%	35 years
1.5%	47 years
1.0%	70 years

A difference of 1.0% can equal *twice* the number of years to double a population!

Another consideration when calculating doubling time has to do with the fluidity of population growth. While using the Rule of 70 in investments is a fairly safe bet, population growth rates are rarely as predictable and can fluctuate dramatically from year to year. Therefore, using a population's doubling time, while useful, must be done with an understanding that it is just a predictor, not a concrete value.

POPULATION MOMENTUM

While a population will respond to environmental resistance by decreasing growth rates, no population can respond instantaneously. The lag time between the drop in fertility rates and the decline in the RNI is known as **population momentum**.

Let's take the country of Germany as an example. After World War II, Germany's population experienced a slight period of increased fertility, similar to the "Baby Boom" in the United States. However, when Germany's "Baby Boomers" eventually reached reproduction age, they experienced a lower fertility rate than their parents' generation. In fact, the Baby Boomer generation in Germany had a fertility rate lower than 2.1%—the replacement fertility rate (see box 2.1 for a discussion of fertility rates and replacement fertility).

Although the German population is growing less than the replacement rate of 2.1%, the actual number of individuals in the population will not reflect the decreased population growth rate until the Baby Boomer generation begins to die and the total deaths begin to exceed the birth rate. In other words, Germany has population momentum: their population size does not yet reflect the lower fertility rate, but they are gaining momentum toward a time when deaths will exceed births and the population growth rate will decline.

BOX 2.1 Fertility Rates and Replacement Fertility

Fertility rates differ from growth rates in that they refer only to the number of offspring born to a female in their lifetime. If the fertility rate of a country's population is 1.0 births per female, then an *average* of one child is born to each reproducing female in their lifetime. If the fertility rate is 3.0, the females are giving birth to an average of three children in their lifetime.

The *replacement fertility rate* refers to the fertility rate needed to replace each individual in the population from one generation to the next without increasing or decreasing the population size—a RNI of 0.0%. Generally, the replacement fertility rate is 2.1 births per female because at this rate, the reproducing females in the population are producing an average of 2.1 offspring in their lifetime—one child to replace the mother and father each, plus an additional tenth of percent to replace the portion of the population that will not reproduce. When a population is no longer producing enough offspring to replace each individual in the parent population, the population will begin to decline.

In reality, the replacement population rate differs by population. In industrialized countries, the rate holds at approximately 2.1 births per female. However, in poverty-stricken countries where child mortality is greater and a higher percentage of females do not live to reproduce, the replacement rate is usually higher, ranging from about 2.5 to 3.3.

PRACTICE: *GROWTH RATES AND DOUBLING TIME*

Complete the following questions to test your understanding of growth rate and doubling time formulas. You may check with your instructor for the correct answers when you are finished. Remember to find a common sample size (denominator) if the sample sizes do not match up!

You are studying a population of foxes to find the growth rate. Here are your observations after one year:

- There were 20 births per 100 foxes observed

- 30 foxes died per every 100 foxes observed

- 15 foxes immigrated into a sample of 50 foxes

- 5 foxes emigrate out of a sample of 50 foxes

1. What is the number of *incoming* individuals in this population? What is the number of *outgoing* individuals?

2. What is the RNI of this population?

3. What is the PGR of this population?

3. 3. 10%
2. –10%
1. 50; 40

Name:

LAB 2 WRITE-UP: POPULATION GROWTH

Exercise 1: Simulated Population Growth

Objectives:

We will simulate the growth of three populations over a period of 10 years.

1. A population of a typical industrializing nation
2. A population of a nation in which birth control is practiced
3. A nation with zero population growth
4. We will graph the data and interpret growth rates
5. We will predict the future growth of the populations

Method:

Each group will be given a container with 100 dice. Members of each group will follow the instructions for the three simulations.

Simulation A:

1. Place 20 dice in your container. Keep the rest of the dice separate as your stockpile.
2. Record 20 as the population size for year 1 of the study.
3. Shake the container while holding it upright, and empty all of the dice onto the lab table.
4. The number on the die determines what should be done with that die.
5. Count the total number of dice now on your lab table, and record this number in your data table in the same row as for other data for that year.
6. Return all dice in your current population to the container.
7. Repeat steps 3-6 for 9 more years.

Number on Die	Results
1	Individual has died.
	Remove die and return to stockpile.
	Record deaths in data table
2 or 3	Females have borne a child.
	Add a die from stockpile to container.
	Record births in data table.
4–6	Males and/or females who have not produced a child

Simulation B:

Follow directions for Simulation A with the following exceptions:

Number on Die	Results
1	Individual has died.
	Remove die and return to stockpile.
	Record deaths in data table.
2	Females have borne a child.
	Add a die from stockpile to container.
	Record births in data table
3	50% have successfully practiced birth control.
	Count the number of die and divide by two. If there is an odd number of "3" dice, round up 1.
	Add this value to your total population.
	Record births in data table.
4–6	Males and/or females who have not produced a child.

Simulation C

Follow the directions for Simulation A with the following exceptions:

Number on Die	Results
1	Individual has died.
	Remove die and return to stockpile.
	Record deaths in data table.
2	Females have borne a child.
	Add a die from stockpile to container.
	Record births in data table.
3	Female has successfully practiced birth control.
4–6	Males and/or females who have not produced a child

Results:

1. Record your results for the simulations in the table below.

Simulation A:

Year	Number of Births	Number of Deaths	Population Size
1	X	X	20
2			
3			
4			
5			
6			
7			
8			
9			
10			

Simulation B:

Year	Number of Births	Number of Deaths	Population Size
1	X	X	20
2			
3			
4			
5			
6			
7			
8			
9			
10			

Simulation C:

Year	Number of Births	Number of Deaths	Population Size
1	X	X	20
2			
3			
4			
5			
6			
7			
8			
9			
10			

2. Using the data from year 10 of each of the simulations, calculate the Growth Rate and Doubling Time for each of the populations. Show work to get credit. Provide an example of a country that has a growth rate similar to each of the simulated populations.

Simulation A (**1 point**):

Simulation B (**1 point**):

Simulation C (**1 point**):

3. Estimating Future Populations:

Using Excel on the lab computers, plot the three simulated populations.

Step 1: Add the data from each simulation into the spreadsheet (where year is on the x axis and population is on the y axis).

Step 2: Highlight the data, go to the insert tab > insert chart > insert scatter plot.

Step 3: Select the data points on the chart. Go to the "Chart Design" tab > "Add Chart Element" > "Trendline" > "Linear".

Step 4: Double click the linear trend line on the graph. On the far right of the screen, select the "Discplay Equation on chart" option at the bottom of the menu.

Step 5: Use the new question displayed on the graph to fill out the table below by plugging in the year for x.

Step 6: Repeat for simulations B & C.

Simulation A (1 point): Simulation B (1 point): Simulation C (1 point):

A. Trendline Equation_____

Estimated Population in 20 years ____

Estimated Population in 50 years ____

Estimated Population in 100 years ___

B. Trendline Equation_____

Estimated Population in 20 years ____

Estimated Population in 50 years ____

Estimated Population in 100 years ___

C. Trendline Equation_____

Estimated Population in 20 years ____

Estimated Population in 50 years ____

Estimated Population in 100 years ___

4. Explain **three** negative environmental impacts that are caused by an increased human population size (**1 point**):

Lab III

Food

Introduction

One of the main concerns with a growing human population is finding ways to produce the extra food to feed so many people. Throughout history, major technological advances have usually preceded an increase in food production. A common thread among industrializing countries is an initial agricultural revolution whereby fewer farmers can produce more food. The people who were once farmers are then free to work in the new economic sectors, such as manufacturing and service industries.

A vast library of terminology accompanies the study of society and agriculture. These terms can be both numerous and confusing, even to experts, but we will focus on only a few select concepts in this lab.

FARMING

According to the U.S. Environmental Protection Agency's (USEPA) Agriculture Department, a single farmer in the United States feeds approximately 130 people. In 2009, less than 1% of the U.S. population classified their occupation as farmer.[1] Farming in the United States is often characterized by large fields of single crops that require massive inputs of inorganic fertilizers, pesticides, and irrigation water. Crops have also been modified, either through genetic engineering or selective breeding, to produce larger yields or to resist pests and herbicides. These technologies have allowed farmers to support billions of people globally, but most of the advancements have come at a cost to our natural environment.

Monoculture

A **monoculture** is the repeated production of a single crop in a large agricultural area. While this industrial agricultural production system is extremely productive, it results in long-term soil degradation and increased susceptibility to disease and pests due to the inherent lack of biodiversity.

Some environmental scientists advocate for a more sustainable alternative to conventional agricultural production, known as **permaculture**. The term permaculture broadly refers to a school of environmental thought, but in the context of agriculture, it especially encourages the production of a variety of crops within a single area. For example, a farmer might choose to grow biofuel crops as groundcover beneath fruit trees. The permaculture approach works to mirror the natural ecology by promoting biodiversity and by minimizing inputs of fertilizers, pesticides, and water.

---

[1] The National Agriculture Compliance Assistance Center. 2009. United States Environmental Protection Agency, vol. 2012.

Fertilizers

The excessive use of agricultural fertilizers can inadvertently cause the degradation of nearby water bodies. Nitrogen and phosphorus (two of the main components of inorganic fertilizers) are required for plant growth. While the addition of fertilizers increases crop yields, excess amounts of nitrogen and phosphorus can run off from agricultural fields and make their way to nearby surface waters. **Eutrophication**, from the Greek word *eutrophia*, means the over enrichment of water. In aquatic ecosystems, nitrogen and phosphorus are **limiting nutrients**, or nutrients that exist in short supply and limit population growth through density-dependent environmental resistance. If excess nitrogen or phosphorus from human sources makes its way into a water body, it may cause algal populations to grow exponentially and cover the water body in algal mats. Eventually, the excess nutrients will be used up and the algae population will begin to die off. The die-off of the algae will result in a high level of organic matter that will go through decomposition. The aerobic bacteria and microbes that decompose this organic matter couple with the physical decomposition process to ultimately reduce or even eliminate the amount of dissolved oxygen in the water column. This results in the death of other aquatic organisms, such as fish.

Genetically Modified Organisms (GMOs)

Crops have been genetically engineered through selective breeding since ancient humans first adopted agriculture. However, genetic engineering in the laboratory has allowed scientists to quickly engineer seeds that result in crops with larger yields or favorable traits, such as pest resistance, weed-killer resistance, or high nutrient content. Most of the monoculture crops in the United States are grown using these seeds, which are collectively known as **genetically modified organisms (GMOs)**.

Despite the potential benefits of GMOs, there are a number of concerns about the use of these crops. For example, some health activist groups are worried that consuming GMO foods may trigger food allergies in sensitive groups of people. There is also some speculation as to whether GMO crops may accidentally crossbreed with other plants and even produce herbicide resistant "super weeds." However, there is still much uncertainty surrounding these questions, as little in-depth scientific research has been conducted on the health and environmental effects of GMOs. The issue remains a hot topic in many western countries.

MEAT PRODUCTION

Similar to agricultural farming, meat production systems and animal feeding operations have become increasingly industrialized in the United States. A **Concentrated Animal Feeding Operation (CAFO)** is a facility where large numbers of animals are raised in confined areas. The intense production of animals results in the accumulation of wastes from manure, urine, and dead animals in a relatively small area of land. This large-scale production of meat impacts human health and the environment.

Antibiotic Resistant Bacteria

Large amounts of antibiotics are added to animal feed to increase the growth of animals. In fact, about 70% of antibiotics in the United States are used on healthy farm animals. The addition of antibiotics increases animal growth by allowing more energy to be allocated to growth rather than to fighting the abundance of pathogens in these unhealthy CAFO environments.[2] Residual amounts of antibiotics are present in the meat humans consume.[3] The overuse of these antibiotics can result in multi-resistant strains of bacteria that are more hazardous to human health than the average disease-carrying microbe.

---

[2] The National Agriculture Compliance Assistance Center. 2009. United States Environmental Protection Agency, vol. 2012.

[3] *A River of Waste: The Hazardous Truth about Factory Farms.* DVD. Directed by Don McCorkell. Cinema Libre Studio, 2009.

Animal Waste

Vast amounts of nitrogen- and phosphorus-rich animal waste are produced in CAFOs. This waste can make its way to nearby surface waters though improper disposal or runoff from rain events and further exacerbate eutrophication problems.

"ORGANIC": WHAT IS IT?

An increasingly popular consumer trend in industrialized countries attempts to address many of the issues discussed above by eliminating some of the artificial processes in modern agriculture. The term "organic" is sometimes used very broadly, but it has a more strict set of requirements under the United States Department of Agriculture (USDA). A "USDA Certified Organic" label may be issued to foods that are grown without the use of synthetic pesticides, herbicides, human sludge fertilizer, and are free of GMO content. Organic meat and poultry comes from animals that are fed organic diets and do not receive antibiotics or growth-promoting hormones. Sometimes (but not always), organic meat and poultry may also refer to animals that have been allowed free range of pastures on which to graze. It is important to do some of your own research into the brands of food you purchase if you are concerned about specific issues—you can't always trust the label!

Organic farming may offer many potential benefits, but only if the correct methods are adopted. In some cases the farmers may be using more traditional tilling methods (like those found on nonorganic farms) instead of methods aimed at soil conservation. In other cases, they may be applying very large amounts of non-synthetic (but still toxic) pesticides and herbicides. Every element of a farming operation must be examined to determine its true cost or benefit to the environment.

Name:

LAB 3 WRITE-UP: FOOD

Watch "Jimmy's GM Food Fight" (BBC Horizon, 2008) and answer the following questions as you watch the documentary. The documentary can be found online at: http://www.documentarytube.com/videos/jimmys-gm-food-fight

1. What is the difference between unmodified soybeans and the GM soybean grown in Argentina?
   (0.5 point)

2. Why are farmers growing the soybeans at such large scales?
   (0.5 point)

3. What is an unfortunate result of requiring enough land for the fields?
   (0.5 point)

4. How do GM soybeans become part of people's meals when they do not eat they soybeans?
   (0.5 point)

5. What is a reason why people avoid eating GM foods?
   (0.5 point)

6. Why are many farmers in Europe not willing to grow GM crops?
   (0.5 point)

7. Can the vegetables we eat be able to survive in the wild? Why?
   (0.5 point)

8. What types of genes are typically placed in GM plants?
   (0.5 point)

9. Besides combating environmental stressors, what can GM plants provide for humans?
(0.5 point)

10. Why are the GM tomatoes not available in European supermarkets?
(0.5 point)

11. How long have GM crops been grown and consumed in the US?
(0.5 point)

12. Why do the Amish choose to grow GM crops?
(0.5 point)

13. What does the BT corn prevent?
(0.5 point)

14. How does the GM cotton help the environment and farmers?
(0.5 point)

15. How can GM plant genes spread to other plants?
(0.5 point)

16. Are GM foods labeled in the US? (Y/N) Is it easy to trace GM free foods? (Y/N)
(0.5 point)

17. How can GM crops help developing nations, such as those in Africa?
    (0.5 point)

18. What implications does the spread of the Black Sigatoka disease have to the crop and to the livelihood of the farmer?
    (0.5 point)

19. What plant genes are inserted into the GM banana crops to help prevent disease? Why?
    (0.5 point)

20. What are YOUR views of GM crops for both developed and developing worlds?
    (0.5 point)

$\mathcal{L}ab$ IV

Energy

Introduction

Based on known reserves and current rates of consumption, we may have less than 50 years of oil supplies left.[1] Alternative energy sources will thus play an important role in our nation's energy future. In this lab, you will learn about the challenges and benefits associated with a number of different energy sources.

BACKGROUND

Different energy sources may be used for different purposes, including electricity generation, transportation, and heating. In the United States, the greatest energy demand is for electricity. Most of our electricity is generated in large power plants, with a total of 6,600 power plants in operation today. In this lab, we will focus primarily on the energy sources used to produce electricity.

Generally speaking, power plants use a source of energy (e.g., coal and uranium) to boil water to generate steam to turn a turbine (Figure 4.1). Electricity can be generated from a variety of sources. Fossil fuels (coal, natural gas, and petroleum) are the most common source, accounting for 66% of U.S. electricity generation in 2017.[2] Nuclear sources are also important, contributing 20% of all energy generation. Renewable energy sources, such as wind, solar, hydro, biomass, and geothermal energy, make up only 12% of our electricity portfolio (Figure 4.2). The generation methods used with many renewable sources differ from the methods used in traditional power plants, like the one pictured in Figure 4.1.

Transportation is the second largest demand sector, with the bulk of this energy coming from petroleum. There is an increasing amount of interest in the use of natural gas as a transportation fuel source due to concerns about the limited availability of oil. Because the United States has an abundant supply of natural gas, some advocate for the U.S. to increase our energy independence and reduce dependence on foreign nations for oil by transitioning to this fuel source within our own borders. However, critics of this policy point out that natural gas is still a limited resource that results in greenhouse gas emissions, and they argue that we should opt for *energy security* over *energy independence* by switching our focus to renewable energy, which may be less prone to international political instability and market failure.

Scalability

Scalability is an important test to consider for any energy source. To be considered scalable, sources must be able to keep up with the rapidly increasing demand for energy. As you examine the impacts of each energy source, keep in mind how effectively it can be used to meet a growing population's energy needs.

---

[1] BP p.l.c. 2014. "BP Statistical Review shows strength of global energy system amid disruptions and shifting world economy." BP Press Release, June 16. http://www.bp.com/en/global/corporate/press/press-releases/bp-statistical-review-shows-strength-of-global-energy-system-ami.html.

[2] U.S. Energy Information Administration, 2013. *Electrical Power Monthly*.

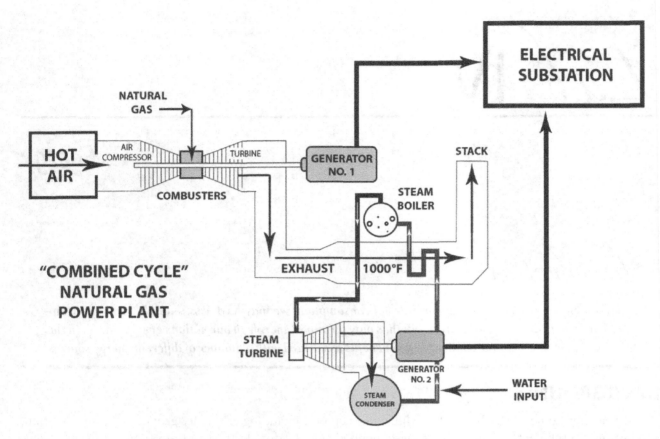

Figure 4.1. *In a traditional power plant, a source of energy is used to boil water to generate steam to turn a turbine. In the "combined cycle" natural gas power plant shown above, the combustion of natural gas releases hot gases used to push a turbine and generate electricity. Waste heat from this first process is used in a second process to heat water and produce steam, which is also used to turn a turbine and generate electricity.*

Figure 4.2. *Energy sources used for U.S. electricity generation in 2017.*
Source: U.S. Energy Information Administration.

Name:

Partners:

Exercise 1: Energy Sources (3 points total)

This exercise will be done in class. You will be divided into groups and assigned an energy source. Use your lab manual or online resources (e.g. www.seedsenergy.ca, ww.energy.gov) to research your group's energy source (use the wind row as an example). The groups will then present their findings to the class, and we will compile a "master table" with this information. You may either paraphrase the "master table" into your write-up or fill it in yourself. WHEN YOU SUBMIT THIS ASSIGNMENT, YOU MUST HAVE ALL THE ROWS FILLED IN.

Please note that petroleum includes both oil and natural gas. In this chart, petroleum is separated into two rows to more accurately distinguish their advantages, disadvantages, and use.

Energy Source	Nonrenewable or renewable	Where does the energy originate?	What is this energy used for (e.g. electricity generation, transportation, heating) and how is this done?	Advantages of using this source	Disadvantages of using this source
Wind	renewable	Unequal solar warming of Earth's surface and atmosphere produces wind	Electricity generation- Wind turns turbine blades and via an axel turns a generator	Many wind places, relatively cheap, no emissions	Intermittent, possible hazard to birds and bats, noise, light flickering
Petroleum Natural Gas					

Energy Source	Nonrenewable or renewable	Where does the energy originate?	What is this energy used for and how is this done?	Advantages of using this source	Disadvantages of using this source
Nuclear					
Biomass					
Coal					

Energy Source	Nonrenewable or renewable	Where does the energy originate?	What is this energy used for and how is this done?	Advantages of using this source	Disadvantages of using this source
Petroleum Oil					
Tidal & Wave					
Solar					

Energy Source	Nonrenewable or renewable	Where does the energy originate?	What is this energy used for and how is this done?	Advantages of using this source	Disadvantages of using this source
Hydro					
Geothermal					

1 http://www.eesi.org/fossil_fuel

Exercise 2: Electricity in Your Region (4 points total)

This exercise is to be done individually. Choose 1–3 of the energy sources from the chart you completed in Exercise 1. In a 300–500 word essay, provide an argument about the benefits of the energy source(s) and why you think their use should be expanded or continued in your region (you can use north Texas or your hometown, just distinguish which one you chose). Be sure to explain why the energy source(s) would be both available and feasible for your region specifically (i.e., if you chose north Texas, do not choose tidal energy). CITE YOUR SOURCES BY PROVIDING THE URL. Attach the completed essay to this write-up.

Exercise 3: Natural Gas (3 points total)

Go to http://exploreshale.org/ This website looks at the Marcellus Shale in Pennsylvania. Familiarize yourself with the website, being sure to scroll all the way down and to the right and left. Answer the following questions in your own words. Copy and pasting will result in a zero for the whole write-up.

1. What is shale? How is shale different from other rocks? (**0.25 point**)

2. Where is the Marcellus Shale located? How large is the Marcellus Shale? (**0.25 point**)

3. How did the Marcellus Shale form? (**0.5 point**)

4. What is the typical natural gas well depth? How are natural gas oil wells constructed? (**0.5 point**)

5.   What is fracking? What are the stages of fracturing and production? Describe each step? **(0.5 point)**

6.   How much water is used in hydraulic fracturing? What else is used in frack fluid? Why are proppants used? **(0.5 point)**

7.   What happens to the frack water after it's done? Can water be contaminated? **(0.5 point)**

Lab V

The Greenhouse Gas Effect and Carbon Budgets

Introduction

In 2007, a consortium of college and university presidents concerned by the potential impacts of climate change agreed to sign an agreement to achieve climate neutrality by reducing their greenhouse gas emissions. Since 2007, hundreds of colleges and universities have signed the agreement. In fact, it's very likely that your own school is a member! Part of the commitment is the inventory of the greenhouse gases emitted by the university or college. In this lab, you will determine the amount of carbon your campus emits in the form of carbon dioxide and the amount it stores as biomass in trees.

BACKGROUND

Greenhouse Gases—What Are They?

The term "greenhouse gases" (GHGs) is actually a collective term for a group of different chemical compounds. Table 5.1 lists several of the most important GHGs.

GHGs, like all chemical compounds, consist of atoms that are hooked to each other through chemical bonds. The chemical bonds vibrate when they are hit by energy, particularly infrared energy. This is not a form of energy human eyes can see, like the waves in the visible light spectrum, but we actually *feel* it—as heat! **Infrared radiation** is invisible thermal energy that can be absorbed and re-emitted by greenhouse gases.

The chemical bonds in GHGs vibrate during infrared energy absorption because the energy excites the chemical bonds and make the compounds move around faster, increasing the temperature of the surrounding air. This interaction between infrared energy and greenhouse gases in the atmosphere is central to the *greenhouse effect*.

Table 5.1. *Sources of many important greenhouse gases.*

GHG	Human Sources
Carbon dioxide	Burning fossil fuels, deforestation
Methane	Feedlots, rice fields
Nitrous oxide	Vehicles, machinery, agricultural fields
Chlorofluorocarbons	Coolants, spray cans
Water vapor	Evapotranspiration (natural)

BOX 5.1   The Light Spectrum

The electromagnetic light spectrum consists of waves that carry electricity and magnetism. Visible light is just a very small part of the light spectrum. It is also the only part of the spectrum we can see with our eyes. Bees can see some ultraviolet (UV) light because the flowers they pollinate produce special pigments in the UV range. UV, visible, and infrared light are all simply different forms of radiation.

Electromagnetic waves travel through space with a specific frequency (h) and speed (the speed of light, or 300,000 km/second). Frequency refers to the number of waves cycling past a fixed point and is expressed in units of cycles per second (cycles/second, or hertz). We can also measure waves' properties in terms of wavelength. Wavelength (λ) is the distance from peak to peak between the cycles of the wave, or the distance between the troughs in the waves. Wavelength and frequency are inversely related to each other. As wavelength goes up, frequency goes down; as wavelength goes down, frequency goes up.

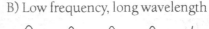
A) High frequency, short wavelength   B) Low frequency, long wavelength

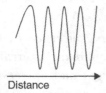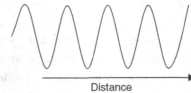

Distance   Distance

The energy content in light waves is described by the following equation:

$$E = h \times f$$

where E is energy expressed in units of joules, h is a constant called Planck's constant in units of joules per hertz, and f is the frequency of light in units of hertz. This equation tells us that energy transmitted with light waves is directly proportional to the frequency of light. If f *increases*, E increases. Alternatively, if f decreases, E will decrease. The energy level of different forms of radiation is important when it comes to the effect of radiation of humans. For example, UV radiation can cause sunburns and even skin cancer because it is *high energy*. You would not want to go through an X-ray machine without protection because it is also very high energy, and could damage the cells inside your body.

Because infrared light is low frequency, it is also low energy. Infrared waves carry energy that is detected as heat. Night-vision goggles have semiconductor detectors that detect the heat released by the human body in the infrared wavelengths (or frequencies).

The Greenhouse Effect

The **greenhouse effect** describes the rise in the earth's average global temperature due to an increase in infrared energy-absorbing gases in the lower atmosphere. This section discusses each step in the greenhouse effect process.

The sun emits many kinds of light waves—nearly every wavelength across the electromagnetic spectrum (see Box 4.1). Most of these waves are directed away from the earth by the earth's magnetic field before they can reach the planet. The majority of light waves that make it to the earth are in the visible and ultraviolet (UV) range. Most UV waves are absorbed by the protective ozone layer in the stratosphere (in the upper atmosphere), leaving visible light as the primary waves that reach the surface of the earth.

There are two things that can happen to light (energy) when it strikes matter:

1.   **Reflection**—The light waves simply "bounce" off the matter, and the energy/frequency of the light does not change.

2. **Absorption**—The light (energy) is absorbed by matter. The chemical bonds in the molecules that compose the matter are consequently excited and the object is heated. *The absorption process is immediately followed by* **re-emission** *(also known as re-radiation), where the heated object then radiates infrared light energy (heat) in all directions.*

After radiation from the sun enters earth's atmosphere, the following steps occur (also see Figure 5.1):

1. Some radiation is reflected from the earth's surface back into space, but some radiation is absorbed on the surface. Snow and ice are good *reflectors* of radiation, while soil, water, and vegetation are generally good *absorbers*.

2. Absorption of energy causes the electrons in the absorbing matter to move faster, heating the air around it. The heated earth radiates that thermal energy in all directions in the form of infrared radiation.

3. Some of the re-emitted infrared radiation from the earth's warmed surface is directed toward the atmosphere. Remember that GHGs are good absorbers of low-energy radiation, including infrared radiation. Some of the radiation passes through the earth's atmosphere into space, but some radiation is absorbed by GHGs in the lower atmosphere, exciting electrons in GHG molecules and generating more infrared energy.

4. The excited GHG molecules re-radiate infrared energy in all directions, warming the lower atmosphere and the earth's surface.

5. The re-radiated infrared that reaches the surface of the earth will warm the earth even more, and the cycle will begin again.

The greenhouse effect is a vital element in making the earth's climate warm enough for life to exist. However, a dramatic increase in the concentration of GHGs in the atmosphere could alter the climate in ways for which current plant and animal life on earth is not adapted. Global warming may also put stress on many important relationships between the natural environment and human civilization, including water supply and agriculture.

The increase in GHGs in the atmosphere has been a leading cause since the dawn of the Industrial Revolution in the eighteenth century, when mankind rapidly expanded energy-consuming processes and began to depend

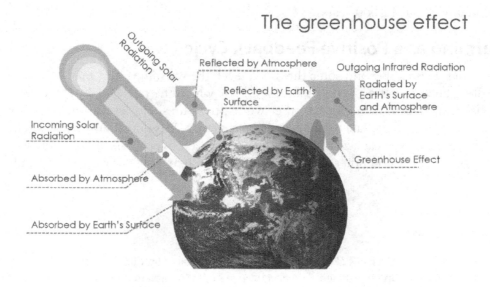

Figure 5.1. *Most of the energy reaching the surface of the earth is in the form of short, visible waves; the absorbed waves are re-emitted as infrared radiation, which is absorbed and re-emitted once more by GHGs in the lower atmosphere. The effect is a general warming of the earth's surface and lower atmosphere.*
Source: Slattery, M.C. (2020). Contemporary Environmental Issues, 6th Edition. Kendall Hunt.

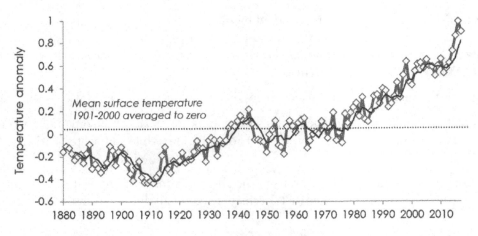

Figure 5.2. *Observed globally averaged combined land ocean surface temperature anomaly, 1850–2017. The annual average temperatures have increased more during each of the past three decades than they did in all preceding decades since 1850 combined.*
Source: www.ncdc.noaa.gov/cmb-faq/anomalies.php

on fossil fuels as the primary source of energy.[1] The Intergovernmental Panel on Climate Change (IPCC) reviewed thousands of scientific publications and concluded that:

> *"Global atmospheric concentrations of CO_2, methane (CH_4), and nitrous oxide (N_2O) have increased markedly as a result of human activities since 1750 and now far exceed pre-industrial values determined from ice cores spanning many thousands of years."*

> —Intergovernmental Panel on Climate Change, Fourth Assessment Report, 2007.[2]

Furthermore, the global average temperatures for each decade since 1980 have been successively warmer than each preceding decade since 1850 (see Figure 5.2), and concentrations of GHGs in the atmosphere are the highest they have been in approximately 800,000 years. To put this in perspective, the anatomically modern *Homo sapiens* (the human species) has only been around for approximately 200,000 years, and the practice of agriculture began only about 12,000 years ago!

Global Warming as a Positive Feedback Cycle

Once Earth becomes warmed beyond some threshold level, the warming effect may become amplified into a "runaway heating effect," or a positive feedback cycle by which the rate of the earth's warming increases as average global temperature increases. In other words, this threshold level is effectively a "point of no return," after which it may be too late to mitigate climate change. It is uncertain exactly what this threshold level is, or if we have already reached it. This positive feedback hypothesis is based on scientists' understanding of historical climates on other planets in our solar system, some of which have undergone such a runaway heating effect.

---

[1] Intergovernmental Panel on Climate Change. 2013. "Summary for Policymakers." In: Climate Change 2013: The Physical Science Basis. Contribution of Working Group I to the Fifth Assessment Report of the Intergovernmental Panel on Climate Change. Stocker, T.F., D. Qin, G.K. Plattner, M. Tignor, S.K. Allen, J. Boschung, A. Nauels, Y. Xia, V. Bex and P.M. Midgley (eds.). Cambridge University Press, Cambridge, United Kingdom and New York, NY, USA.

[2] Intergovernmental Panel on Climate Change, 2007. "Summary for Policymakers." In: Climate Change 2007: The Physical Science Basis. Contribution of Working Group I to the Fourth Assessment Report of the Intergovernmental Panel on Climate Change. Solomon, S., D. Qin, M. Manning, Z. Chen, M. Marquis, K.B. Averyt, M. Tignor and H.L. Miller (eds.) Cambridge University Press, Cambridge, United Kingdom and New York, NY, USA.

Climate scientists are concerned about three main mechanisms that may cause a positive feedback warming cycle:

1. **Methane release from permafrost**— A massive amount of the greenhouse gas methane (CH_4) exists locked in the permafrost (soil that is frozen year-round) in very cold areas like Northern Canada and Siberia. This methane, which may be up to 20 times more potent than CO_2 as a greenhouse gas, is beginning to be released now that the earth has warmed enough for the permafrost to thaw—a geologic event that last occurred on a large scale about 11,000 years ago.[3] When this methane is released, it acts with other GHGs to warm the earth further, which in turn melts more permafrost and releases more methane. The warming process increases in speed with each turn of the positive feedback cycle.

2. **Atmospheric water**—Water is always present in the atmosphere as vapor. The amount of water vapor in the atmosphere at any given moment in time is a function of temperature: the warmer the air, the greater the evaporation rate and, therefore, the more water vapor in the air. This might not seem important until we remember that water is also a greenhouse gas! The warmer Earth becomes, the more atmospheric water will be generated, and the more water vapor in the atmosphere, the warmer the earth will become.

3. **Melting of the polar ice caps**—Snow and ice are excellent reflectors of radiation. Polar ice caps cover a significant portion of the globe, which means that a significant portion of incoming radiation is reflected back into outer space. However, because of recent warming trends, the polar ice caps are melting. While ice reflects solar radiation, water absorbs solar radiation and re-emits infrared radiation back toward the atmosphere. Therefore, as the surface area of the ice caps becomes smaller, the amount of radiation absorbed by the earth exponentially increases and creates the positive feedback cycle.

CARBON BUDGETS

Carbon dioxide (CO_2) is a greenhouse gas that plays a large role in global climate change. The carbon atom in CO_2 is important because plants and other primary producers use it to create carbohydrates, which are the most important source of energy for most organisms. This process is called **photosynthesis**, which is the biochemical process by which primary producers use CO_2 and water to generate carbohydrates in the presence of sunlight. The equation for photosynthesis is given below in Eq. 4.1.

$$6\,CO_2 + 12\,H_2O \rightarrow C_6H_{12}O_6 + 6\,O_2 + 6\,H_2O \tag{4.1}$$

$C_6H_{12}O_6$ is glucose, the most basic carbohydrate. These carbohydrates can be used for energy, or be transformed into plant structures like leaves, bark, and roots. Thus, the plant's biomass can be used as a proxy measure for the rate of carbon assimilation. **Carbon assimilation** describes the process by which primary producers convert atmospheric CO_2 into carbohydrates during photosynthesis.

After primary producers create carbohydrates, higher consumers break them down to create energy. In the process, they release CO_2 in a process called cellular respiration. **Respiration**, given in Eq. 4.2, is the reverse process of photosynthesis.

$$C_6H_{12}O_6 + 6\,O_2 \rightarrow 6\,CO_2 + 6\,H_2O \tag{4.2}$$

In cellular respiration, energy is released when glucose is broken up by oxygen. Although all oxygen breathing species will emit CO_2 and water vapor during respiration, the single greatest source of CO_2 in our atmosphere since the Industrial Revolution is not from respiration from the biosphere, but from the burning of fossil fuels.

---

[3] Walter, K.M., S.A. Zimov, J.P. Chanton, D. Verbyla, and F.S. Chapin III. 2006. "Methane bubbling from Siberian thaw lakes as a positive feedback to global warming." *Nature* 443, 71–75. doi:10.1038/nature05040

Carbon assimilation is important because it *removes* CO_2 from the atmosphere and locks it away in plant structure.

Carbon assimilation occurs anywhere there are photosynthetic organisms. For example, grasses take in carbon dioxide in their blades and store it as biomass, while microscopic organisms in the soil underneath take in oxygen and emit carbon dioxide after respiration. The net amount of carbon emitted subtracted from carbon assimilated is known as **net primary productivity (NPP)**. A given geographic area is a **carbon source** if more CO_2 is being emitted than photosynthetic organisms in the same area can assimilate into carbohydrates—in other words, if the NPP is *negative*. An area is a **carbon sink** if the NPP is *positive*, or if more carbon assimilation is occurring than CO_2 is emitted.

The goal of today's lab is to determine the NPP at your university by developing a carbon budget. A **carbon budget** measures the amount of CO_2 assimilated by primary producers and balances this measurement with the amount of CO_2 emitted in the same area. Today, we will measure the amount of carbon assimilated by trees on campus and compare this to the amount of carbon dioxide emitted by fossil-fuel burning passenger vehicles on campus.

BOX 5.2   Quantifying Carbon Assimilation

Approximately half of a tree's weight is wood, and wood is nothing more than long chains of carbon woven together into strong fibers. Scientists have developed mathematical relationships between the diameter of a tree trunk at breast height and the amount of carbon (biomass, expressed in kilograms of carbon) for several different species of trees.

You can see that as the diameter at breast height (DBH) increases, so does biomass. Different species of trees have different growth forms and wood densities, so each species has its own unique mathematical relationship between DBH and biomass. Notice that the relationship is exponential.

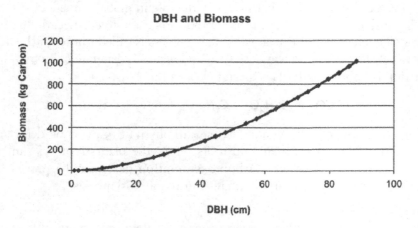

Once the biomass of a tree has been determined, we can use this information to calculate the age of the tree:

$$\text{Age} = (m \times \text{DBH}) + b$$

where *m* is the slope of a line and *b* is the ordinate axis intercept. Each species of tree has its unique values for *m* and *b*. Determining the age of a tree lets us know how long the tree has been building biomass through photosynthesis.

Finally, once we have calculated the biomass and age of the tree, we can divide the biomass by the age to determine how much carbon the tree has assimilated per year or per day during its lifetime.

Name:

LAB 5 WRITE-UP: IS YOUR UNIVERSITY A CARBON SINK OR SOURCE?

Part I: Procedure for Collecting Data (30 minutes) (4 points)

You will be divided into two primary groups: Tree Groups and Car Groups. You will be assigned a specific area of campus to assess.

GROUP_____

AREA_____

Tree Groups (1 point)

Goal: Determine the amount of carbon assimilated by trees on campus.

1. Collect the supplies needed from instructor: tape measure, tree identification sheets.

2. Go to assigned territory, measure the circumference at breast height (cbh) of the trees. The standard breast height in the US is 1.4 m (140 cm) above the ground.

3. Record the circumference and species of each tree that you measured. Attempt to measure as many trees as you can in the time allotted. (If you measure a 30 cm cbh for one live oak and see nine additional live oaks with approximately the same size trunk, then you can record a total of 10 live oaks with a 30 cm cbh.)

Species	Circumference (cm)	Species	Circumference (cm)

Species	Circumference (cm)	Species	Circumference (cm)

Car Groups (1 point)

Goal: Determine the amount of carbon dioxide emitted by passenger vehicles on campus.

1. Locate the cars in your assigned area of campus.

2. Count the number of cars, SUVs, and trucks. Record these numbers before returning to the classroom.

   Cars _____   SUVs _____   Trucks _____

3. When you return to the classroom, visit "http://www.fueleconomy.gov" and click on "Find a Car," then "Find a Car – Home." Select one model of a car, truck, and SUV that is most commonly found in your quadrant. (For example, if you think most cars in the parking lot were Honda Accords, se-lect Honda Accord to represent all cars. If you think most trucks are Ford F-150, select Ford F-150 to represent all trucks, etc.) Fill in your selection under "Browse by Model."

4. Click "Go." Select your car from the list and then click the "Energy and Environment" tab. (Note: If after you select a model, there are different engine types, select an engine type.) Scroll down to "Greenhouse Gas Emissions," and select "US Tons per Year" under units. Record the number under "Regular Gasoline" on the right.

   US Tons per Year Car _____

5.   Repeat step 4 for a model SUV and model truck

US Tons per Year SUV _____

US Tons per Year Truck _____

6.   Go to Part II. Spreadsheet Procedures and complete the instructions for Carbon Emissions.

Part II. Spreadsheet Procedures

Carbon Assimilation (Tree Groups)

Open the appropriate instructor-provided spreadsheet for the tree groups. You will use this to spreadsheet determine how much carbon dioxide is being assimilated from the atmosphere into the vegetation on a daily basis.

First, read the instructions at the top of the spreadsheet. Then take a look at the data table. *Avoid clicking random cells throughout the data table—the spreadsheet can be very sensitive!*. The table is designed to take your inputs in the highlighted cells (cbh) and calculate the age and biomass of each tree recorded in order to determine the grams of carbon assimilated daily by all the trees on your campus.

1.   Input your cbh values into the first column for your first species. Be sure to fill out one cell for each of the trees measured. For example, if you measured ten live oaks with a 30 cm cbh, you should fill out ten cbh cells in the live oaks rows with "30".

2.   Repeat step 1 for each species.

3.   The spreadsheet will calculate the total grams/day of CO_2 assimilated in the trees you counted on campus. The spreadsheet will then translate this into grams/day of CO_2 assimilated throughout your campus based the approximate total number of trees at your university.

4.   Total CO_2 assimilated: _____ **(0.5 point)**

Carbon Emissions (Car Groups)

Open the appropriate instructor-provided spreadsheet for the car counting groups. You will use this to determine how much carbon dioxide is being released the atmosphere from vehicles on campus on an annual basis.

1.   Enter the total number of cars, SUVs, and trucks into their respective cells

2.   Enter the total tons/year from the EPA Fuel Economy website for your model car, SUV, and truck.

3.   This spreadsheet will use the inputted information and the approximate total parking spaces on campus to determine the total CO_2 emitted in grams/day on your campus.

4.   Total CO_2 emitted: _____ **(0.5 point)**

Part III. Calculating the University's Net Primary Productivity

We now have an estimate how much carbon the trees on your campus assimilate through photosynthesis on a daily basis and how much CO_2 vehicles on your campus emit annually. The final step in calculating your university's NPP is putting this information together.

Find the corresponding tree or car group for your assigned area and elect one person to open the final spreadsheet. Type the output information from the previous two spreadsheets into the appropriate cells in this new spreadsheet.

NPP: _____ [1 pt]

Part IV. Is Your University a Carbon Sink or Source?

Individually, compare the amount of carbon emitted to the amount of carbon sequestered. Is the campus a carbon source or a carbon sink? How do we know? If the campus is a carbon source, what practices can we implement to become a sink? If the university is a carbon sink, what practices are we implementing to allow for that to happen? Write a few paragraphs explaining your answers. Give specific examples to back up your position. (Hint: Think back to the ecological footprint lab) (**6 points**)

Lab VI

Island Biogeography and Biodiversity

Introduction

As we learned in Lab II, the human population is continuing to grow at an unprecedented rate. With this growth comes the challenge of finding a balance between our need for more resources and our desire to preserve our remaining natural ecosystems. This pressing issue led to the development of a relatively new branch of ecology, called **urban ecology**. *Urban ecologists study the interactions of organisms within an urban environment and work to incorporate the needs of a human population with the needs of nature.*

Ecologists attempt to measure each species within an ecological community in order to better understand the **biodiversity**, *or biological diversity, of the community. Although biodiversity is a broad concept with many definitions, it can be simplified as meaning variety of life at each level of organization. It could also be described as a measure of how many organisms there are and how different they are from each other.[1] Biodiversity includes the genetic diversity of individuals within a species, the number of different types of species, and the variety of habitats in which these species live.*

MEASURING BIODIVERSITY

One reason scientists study biodiversity is to evaluate the health of an ecosystem. In general, an ecosystem with high biodiversity is a healthier ecosystem (there are some exceptions to this rule; for example, the arctic tundra's low biodiversity does not prevent it from being a healthily functioning ecosystem). Biodiversity plays a key role in maintaining the ecosystem functions that provide humans with a variety of ecosystem goods (i.e., timber or fish) and services (i.e., water purification). Quantifying biodiversity also allows scientists to predict what may happen if a species is removed from its habitat or if a new species is introduced.

There are many methods for assessing the biodiversity of an area. **Species richness** is the most basic measure of biodiversity and is defined as the number of species within a community, landscape, or region. While it is a useful measure, species richness alone cannot provide the complete picture, because it treats rare species and common species as equals *without regard to population size*. For example, if you measure the species richness of an area that is home to hundreds of individuals of one species and only one individual of another species, should you treat it the same as an area that is home to the same two species with an equal number of both those species? Scientists use **species evenness** to measure the relative abundance of different species or how rare or common they are.

Figure 6.1 illustrates the difference between species richness and species evenness.

A biologist is trying to determine the biodiversity of hardwood trees in four different forest study areas. She found that while each area had the same number of tree species, the areas had considerably different numbers

---

[1] Hens, Luc, and Emmanuel K. Boon. 2003. "Causes of biodiversity loss: A human ecological analysis." *MultiCiência* 1. Accessed April 22, 2013. http://www.multiciencia.unicamp.br/artigos_01/A1_HensBoon_ing.PDF.

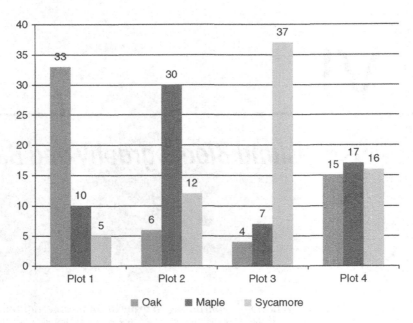

Figure 6.1. *Species counts from four plots of forest.*

of individuals of each tree species. In other words, although the study areas had the same species richness (three species), they were quite different in their compositions. Describing the four areas of forest as the same would be very misleading. By taking into account species evenness, we can demonstrate that these areas are different in composition. When a scientist uses species richness *and* evenness to measure biodiversity, they are measuring the **heterogeneity** of that area. Heterogeneity describes how different one area is from another.

Simpson's Biodiversity Index (D)

To calculate heterogeneity and estimate biodiversity, scientists use a mathematical index. There are a variety of indices available, but we will use *Simpson's Biodiversity Index (D)*. The equation for Simpson's Biodiversity Index is

$$D = \Sigma \left(\frac{n}{N}\right)^2$$

Where:

n = the total number of individuals of a single species

N = the total number of individuals of *all* species combined

---

BOX 6.1   Mathematical Hints

When you are using the equation for Simpson's Biodiversity Index, the Greek symbol Σ simply indicates that you need to calculate the sum of what follows.

Also, don't forget to follow the order of operations when performing your calculations: parentheses, exponent, multiplication/division, then addition/subtraction.

Simpson's Biodiversity Index (D) will yield a number between 0 and 1. With this equation, 0 represents infinite diversity, and 1 represents no diversity. Since this is not very intuitive, we can express D in one of two other ways:

1.   *Simpson's index of diversity* (1−D): The result still falls between 0 and 1, but now 0 represents no diversity, and 1 represents infinite diversity. Note that Simpson's index of diversity is *not* the same as Simpson's Biodiversity Index.

2. ***Simpson's reciprocal index*** $\left(\dfrac{1}{D}\right)$: This equation yields a number greater than or equal to 1 but has no upper limit; thus, the higher the diversity, the greater the value.

Box 6.2 shows the index calculations for each of the four study areas of forest shown in Figure 6.1. As you can see, even though each study area has the same number of trees and the same species of trees, area 4 has the highest heterogeneity and area 3 has the lowest heterogeneity.

Box 6.2: Biodiversity Index Practice

Let's use Simpson's Biodiversity Index to find the index values for the biologist's forest plots shown in Figure 6.1:

Plot 1

$n_a = 33$

$n_b = 10$

$n_c = 5$

$N = 33 + 10 + 5 = 48$

$D = \left(\dfrac{33}{48}\right)^2 + \left(\dfrac{10}{48}\right)^2 + \left(\dfrac{5}{48}\right)^2 = .53$

$1 - D = 1 - .53 = .47$

$1/D = 1/.53 = 1.89$

← First, assign values to each variable. Since there are three tree species in Plot 1, you will have 3 values for n. Here they are designated n_a, n_b, and n_c.

← To find N, add together the total number of individuals of all species.

← Next, plug the values for n and N into the formula for Simpson's Biodiversity Index (D). Then take the sum.

← After you have calculated D, finding $(1 - D)$ and $\left(\dfrac{1}{D}\right)$ is simple.

Below are sample calculations for the other three plots:

Plot 2

$D = \left(\dfrac{6}{48}\right)^2 + \left(\dfrac{30}{48}\right)^2 + \left(\dfrac{12}{48}\right)^2 = .47$ 　　 $1 - D = 1 - .47 = .53$ 　　 $\dfrac{1}{D} = \dfrac{1}{.47} = 2.13$

Plot 3

$D = \left(\dfrac{4}{48}\right)^2 + \left(\dfrac{7}{48}\right)^2 + \left(\dfrac{37}{48}\right)^2 = .62$ 　　 $1 - D = 1 - .62 = .38$ 　　 $\dfrac{1}{D} = \dfrac{1}{.62} = 1.61$

Plot 4

$D = \left(\dfrac{15}{48}\right)^2 + \left(\dfrac{17}{48}\right)^2 + \left(\dfrac{16}{48}\right)^2 = .33$ 　　 $1 - D = 1 - .33 = .67$ 　　 $\dfrac{1}{D} = \dfrac{1}{.33} = 3.03$

ISLAND BIODIVERSITY AND POPULATION ECOLOGY

The biodiversity of an island is very vulnerable to adverse impacts because of its isolation. Imagine that a tornado has wiped out all the trees in your city or town. It would not take long for new tree populations to take hold in the area, because there are probably trees surrounding your city. On the other hand, if all the trees on an island were wiped out, the likelihood that the tree populations would be restored would be slim, since the island is surrounded by water and isolated from a source of trees. This scenario demonstrates why maintaining healthy populations of island species is so critical; if they are not maintained, then they are at high risk of becoming extinct. A **minimum viable population** is needed for all species, meaning that there is a minimum number of individuals needed for the survival of a species. If the population declines below this minimum number, the remaining individuals will no longer have the genetic variability to successfully breed, and the species may go extinct in just a few generations.

Human activities have greatly exacerbated species extinctions on islands. The dodo, an island bird of Mauritius, became extinct when Portuguese sailors overhunted them. In the Galápagos Islands, only 10 of the original 15 subspecies of the giant tortoise remain. Beginning in the seventeenth century and continuing until the twentieth century, whalers, pirates, and other seafarers captured hundreds of thousands of the giant tortoises to eat on long voyages. Early settlers to the islands also used the tortoises as a source of meat and oil. In more recent decades poachers have also killed tortoises, even though they are a legally protected species. These examples demonstrate why losses of island species comprise 80% of all recorded extinctions.

Not all islands are equally diverse. For example, Papua New Guinea (PNG) is a collection of several islands just north of Australia and contains nearly seven percent of the world's biodiversity in just one percent of their land mass. PNG is known for its spectacular collection of rare and endemic bird species. Scientists estimate that the islands are home to a total of 740 species of birds, of which 77 species are considered endemic (not found elsewhere) to the mainland and surrounding islands that make up PNG. The central region of the mainland is a hot spot for many species of the rare and endemic birds-of-paradise. Although one might guess that the same species of birds-of-paradise could be found on each island in PNG, the species are not equally distributed. For instance, the Goldie's bird-of-paradise is only found on two of the Trobriand Islands, located just off of the southern peninsula of the mainland.[2]

Scientists have observed similar patterns of surprising differences in biodiversity between geographically close islands. To explain the uneven distribution of species among islands, ecologists developed the **island biogeography theory.** This theory has three main tenets:

1. Immigration rates are a function of the distance of the island from the source area (the mainland).

2. Extinction rates are a function of island area.

3. The species richness of an island is the balance between immigration rates and extinction rates. Over time, the countervailing forces of immigration and extinction come to equilibrium and result in a relatively constant number of species.

Below, the principles of island biogeography are explained graphically.

A) Distance from the mainland influences immigration rates. The more isolated an island is, the lower its rate of immigration. Furthermore, as species richness increases, the rate of immigration drops due to increased competition for resources on the island. There are fewer niches available.

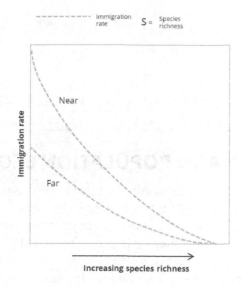

[2] Iamo, Wari, and John Michael. 2010. "Papua New Guinea's Fourth National Report to the Convention on Biological Diversity." Department of Environment and Conservation: Papua New Guinea. Accessed January 15, 2014. http://www.sids2014.org/content/documents/141NBSAP.pdf.

B)   Extinction rates depend on the size of the island. The smaller the island is, the higher its extinction rate, due in part to a lack of resources and less habitat heterogeneity. Additionally, as the species richness of the island increases, extinction rates also increase.

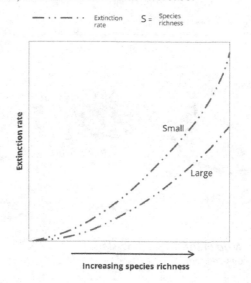

C)   Species richness is the balance between immigration and extinction rates.

One example of island biogeography theory comes from the Galápagos Islands (Figure 6.3). Isabela Island, the largest island in the archipelago, is home to 99 bird species. Pinta Island contains only 54.[3] Both islands are approximately the same distance from the mainland, so island area is most likely a major factor in the difference in bird diversity. According to island biogeography theory, we would expect both islands to have similar immigration rates, but Pinta Island should experience higher extinction rates due to its smaller area.

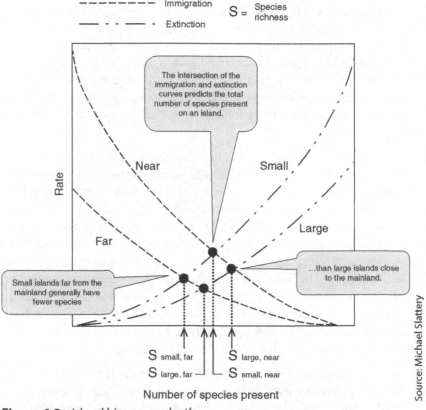

Figure 6.2. *Island biogeography theory curves.*

---

[3] Lepage, Dennis. 2014. *Avibase – The World Bird Database*. Last modified September 30, 2014. http://avibase.bsc-eoc.org/.

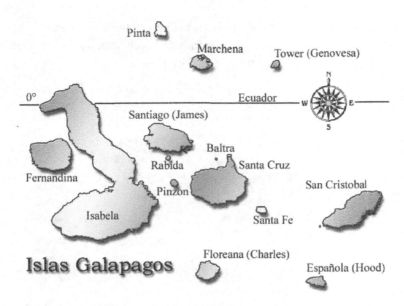

Figure 6.3. *Map of the Galápagos Islands.*
Source: www.ecuaworld.com

Conservation Areas and Islands

Interestingly, island biogeography theory does not apply only to actual islands. It applies to *any habitat that is surrounded by other areas that are unsuitable for the species of that habitat.* For example, the assumptions of island biogeography theory apply to lake environments because they are bodies of water surrounded by land. A forest that is surrounded by human development will follow island biogeography theory as well. Central Park in New York City, as a forest surrounded by an urban area, is a good example of an "island." It has low species richness because it is far from a "mainland," that is, another forest. In addition, any species living within Central Park are vulnerable to extinction because the park functions like a small island. Many conservation areas in the United States face similar problems. Human development surrounds most wilderness areas, leading to **habitat fragmentation**. Roads, agricultural fields, parking lots, homes, and other types of development break up large and continuous patches of habitat into many smaller discontinuous units. Human development and habitat fragmentation transform wilderness and conservation areas into small, isolated islands, in effect. As such, high extinction rates and low immigration rates are a challenge many reserve managers working to protect biodiversity must face.

Conservationists rely on two important strategies to help reduce the vulnerability of animal populations: biological corridors and buffer zones. Each strategy addresses one of the two main factors in island biogeography theory: island size and island isolation (distance from a source of biodiversity). Large islands close to a mainland have the highest biodiversity, and we can imitate these characteristics in conservation areas. First, near islands have higher species richness because of their increased connectivity with the mainland, allowing for easy migration. Conservation areas can mimic this connectivity by creating **biological corridors** (Figure 6.4), which are narrow passes through which species can move from reserve to reserve without crossing an excessively human-dominated landscape. In Southern California, scientists monitored culverts and underpasses to see if animals used them as corridors. Although the underpasses are not areas of natural habitat, the scientists discovered that reptiles, deer, and other small mammals were using them for migration. One of the largest corridors in the world is the Mesoamerica Biological Corridor, which stretches from Mexico to Panama. This corridor has aided in the survival of more than 100 endangered species.

A second tactic used to increase survivability in conservation islands is to make them larger, thus reducing extinction rates. While it is not always feasible to increase the actual size of a conservation area, it may be possible to develop a **buffer zone** around it (Figure 6.5). A buffer zone is a transitional area between a natural environment and human development. Without a buffer zone, human activities may continue up to the

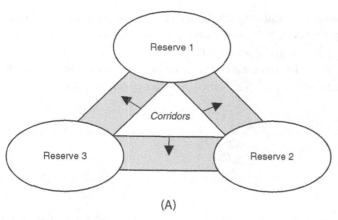

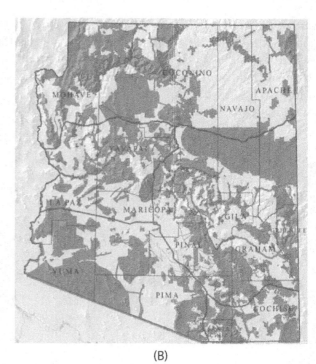

Figure 6.4. *(A) Biological corridors provide escape routes for species to migrate from a mainland (e.g., Reserve 1) to an island (e.g., Reserve 2). (B) Map of proposed wildlife corridors connecting important reserves in Arizona.*
Source: Courtesy of The Nature Conservancy. www.azconservation.org.

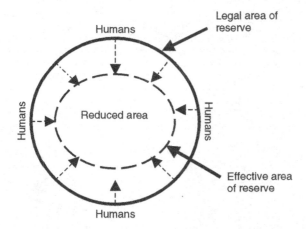

Figure 6.5. *Reserve areas can be effectively reduced by encroaching human impact.*

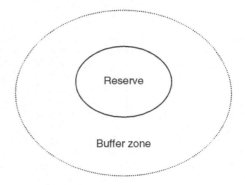

Figure 6.6. *Buffer zones provide protection for nature reserves by limiting humans to sustainable developments in the periphery of reserves.*

boundary of the reserve. When such a sharp edge separates human habitation from a reserve, human influence does not stop at the boundary; in reality, it often extends into the reserve (Figure 6.6). Human activity can severely affect ecological activity, so by restricting human development surrounding conservation areas, human impact on the ecosystems inside is reduced.

Why is it important to conserve wildlife and protect biodiversity? Biodiversity provides a variety of benefits to humans. Many medicines were originally developed from biological sources, and several fabrics are made from plant materials. Biodiversity also contributes to maintaining important **ecosystem goods and services**. Ecosystem goods and services "represent the benefits human populations derive, directly or indirectly, from ecosystem functions."[4] Provisioning services are the most obvious type, and include goods, such as fish, that

---

[4] Costanza, Robert, Ralph d'Arge, Rudolf de Groot, Stephen Farberk, Monica Grasso, Bruce Hannon, Karin Limburg, Shahid Naeem, Robert V. O'Neill, Jose Paruelo, Robert G. Raskin, Paul Sutton, and Marjan van den Belt. 1997. "The Value of the World's Ecosystem Services and Natural Capital." *Nature* 387: 253–260.

are of direct benefit to people. Biodiversity can also help provide regulating services, such as the removal of pollutants from water. Cultural services, such as the aesthetic beauty of an area, and supporting services, which contribute to the functioning of ecosystems in general, are also important. Ultimately, to conserve biodiversity, humans must learn to live more sustainably. How do we conserve wildlife and still meet the needs of our growing population? That is a challenge that we must learn to overcome; otherwise, our wildlife and our way of life may disappear.

Name:

LAB 6 WRITE-UP: BIODIVERSITY AND ISLAND BIOGEOGRAPHY

We will be demonstrating the principles of island biogeography theory using a chain of hypothetical islands as shown in the diagram below. You will conduct two runs. In each run, you will toss seeds from the "mainland" towards the islands in order to simulate colonization by new species.

Once the seeds have been tossed, each group will go to their respective island, separate the beans by type and count them. Record the data of each run in the table provided. **(1 point)**

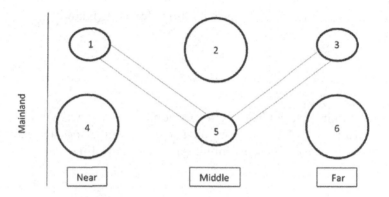

Seed type (on your island)	Run 1: Number of seeds	Run 2: Number of seeds
Species 1 (Pinto--mottled)		
Species 2 (Black)		
Species 3 (Red both shades and sizes)		
Species 4 (Small White)		
Species 5 (Lima – Large White)		

If you were chosen to count seeds in a corridor, please record the results below. Note if you are counting the near or far corridor.

Seed type (in corridor)	Run 1: Number of seeds	Run 2: Number of seeds
Species 1 (Pinto--mottled)		
Species 2 (Black)		
Species 3 (Red both shades and sizes)		
Species 4 (Small White)		
Species 5 (Lima – Large White)		

Calculating Biodiversity

A. Islands without Corridors

1. a) Calculate Simpson's biodiversity index **for your island** for Run 1 and Run 2. The formula is $D=\sum$ where n= the total number of individuals of a species (the number of beans of a certain type) and N= the total number of all the individuals of all the species (all the beans of all types that were on your island). In this equation, 0 represents infinite diversity. **Make sure to show your work. (1 point)**

Run 1:

D =

Run 2:

D =

      b) Calculate the average Simpson's biodiversity index **for your island**.

Average D=

2.   a) Since Simpson's biodiversity index may seem counterintuitive, we will use Simpson's index of diversity (1-D) instead. Just take the D you calculated for your island, and subtract it from 1. Now 0 represents no diversity, and 1 represents infinite diversity. **(1 point)**

Run 1:

1 – D =

Run 2:

1 – D =

      b) Calculate the average Simpson's index of diversity **for your island**.

Average 1 – D

B. Islands with Corridors

Now we are going to look at how biological corridors affect island biodiversity. **Treat islands 1, 3, 5, and the corridors as one large island**. Calculate Simpson's index of diversity (1-D), and include the results in the table below. **Make sure to show your work. (1 point)**

Corridors and Small Islands	Run 1: Number of Seeds	Run 2: Number of Seeds	Simpson's index of diversity (1-D)		Average (1-D)
Islands 1, 3, and 5 and corridors					

Run 1:

D =

1-D=

Run 2:

D =

1 – D =

Analysis of Results

1. How does your island compare? (**1 point**)

2. Did the results of the experiment fit what you would expect based on island biogeography theory? Why or why not? (**1 point**)

A. If your island was close to the mainland, did it have a bigger average index than an island of the same size that was far from the mainland?

B. If your island was far from the mainland, did it have a smaller average index than an island of the same size that was close to the mainland?

C. If your island was in the middle, how did its average index compare to islands of the same size that were closer or farther to the mainland?

D. If your island was large, did it have a bigger index than a smaller island of the same distance from the mainland?

E. If your island was small, did it have a smaller index than a larger island of the same distance from the mainland?

3. What is the effect of the addition of the biological corridor? **(1 point)**

4. How does the index of diversity of 1, 3, and 5 corridors compare to the near large island? The small islands? **(1 point)**

5. What are the potential sources of error for this experiment? In your opinion, did they affect your results? If so, how? If your results did not match your hypotheses, which (if any) of these potential sources of error is the cause? **(1 point)**

Lab VII

Air Pollution, Trophic Levels, and Bioaccumulation

Introduction

From our first breath to our last, our lungs work hard to filter oxygen from the air we breathe. Thus, air quality and air pollution should be something we are very concerned about. Even so, do you know what it means when you hear that it is a "red-level ozone day?" More than likely, it's a part of the local weather forecast that you just don't give much attention. But if you have asthma or another respiratory illness, ozone levels can dictate whether you can afford to go outside that day.

In addition, not all air contaminants are considered harmful while they are suspended in the atmosphere. Some of these pollutants can fall down to Earth's surface through precipitation, settling both into terrestrial and aquatic ecosystems, where they transform into dangerous contaminants for human health. As these compounds pass through the food chain, they accumulate in the tissue of organisms. Since humans are at the top of the food chain, we often ingest large quantities of these contaminants, which in turn may cause severe health concerns, such as neurological damage, toxicity, and even death.

Air pollution is often broken into two main categories: indoor pollution and outdoor pollution. In this lab, you will learn about the consequences to human health from both types.

AIR POLLUTION

Outdoor Air Pollution

While industrialization has led to many economic, social, and technological advances for citizens of the developed world, it does come at a high price. One price that we all pay is in the quality of our air: the cars that we drive, the factories that give us goods, and the electricity that powers our homes all contribute to poorer air quality. Outdoor air pollution comes from both **stationary sources**, such as specific factories, and **mobile sources**, such as the mass of cars on a highway during the morning rush hour (see Box 7.1). In the United States, acceptable outdoor air pollution levels (National Ambient Air Quality Standards, or *NAAQS*) are set by the Environmental Protection Agency (EPA) and usually monitored by each state's environmental agency.

BOX 1.1　Key Terms

Point-source pollution (PS) is a type of pollution where the source can be identified. Poor outdoor air quality is attributed to PS pollution, which arises from smokestacks and car tailpipes.

Nonpoint-source pollution (NPS) is a type of pollution where the source cannot easily be identified. For instance, NPS pollution arises from the movement of water or snowmelt carrying natural and man-made pollutants that runoff from above and below ground. Some of these chemicals come from fertilizers, grease, salt, livestock wastes, or urban runoff. These contaminants then accumulate in larger bodies of water (i.e., lakes, rivers, or oceans). It is considered NPS pollution when you cannot pinpoint single sources such as which farm, power plant, or discharge pipe at which the chemical contamination originated.

There are six principal air quality constituents monitored by state agencies: *ozone, carbon monoxide, lead, nitrogen dioxide, sulfur dioxide, and particulate matter*. The EPA developed the **Air Quality Index (AQI)** to quantify the concentrations of these harmful constituents in our air daily. The AQI expresses concentrations on a scale ranging between 0 and 500. Different ranges within the scale correspond to a color-coded system (see Table 7.1).

With the exception of ozone, all of the NAAQS pollutants are **primary pollutants**, meaning that they are directly emitted into the atmosphere. However, ground-level ozone is a **secondary pollutant**, because it forms as a result of chemical reactions with the primary pollutants *nitrogen oxides (NO_x)* and *volatile organic compounds (VOCs)* in the presence of sunlight (Figure 7.1). In 2016, 122.5 million people lived in counties where air pollution levels *exceeded* NAAQS. Outdoor air pollution impacts children, the elderly, and people

Table 7.1. *The Air Quality Index indicates daily levels of harmful air constituents and the associated health effects. "Sensitive groups" may include children, the elderly, and people with preexisting health conditions, especially respiratory conditions. A higher AQI value indicates more severe health effects for that day. This is why you may commonly hear about orange or even red days in the summer during the morning weather report.*

AIR QUALITY INDEX		
Levels of Health Concern	**Numerical Value**	**Meaning**
GREEN : Good	0–50	Air quality is considered satisfactory, and air pollution poses little to no risk.
YELLOW : Moderate	51–100	Air quality is acceptable; however, for some pollutants there may be a moderate health concern for a very small number of people who are unusually sensitive to air pollution.
ORANGE : Unhealthy for Sensitive Groups	101–150	Members of sensitive groups may experience health effects. The general public is not likely to be affected.
RED : Unhealthy	151–200	Everyone may begin to experience health effects; members of sensitive groups may begin to experience more serious health effects.
PURPLE : Very Unhealthy	201–300	Health alert: everyone may experience more serious health effects.
MAROON : Hazardous	>300	Health warnings of emergency conditions. The entire population is more likely to be affected.

Source: AirNow.org

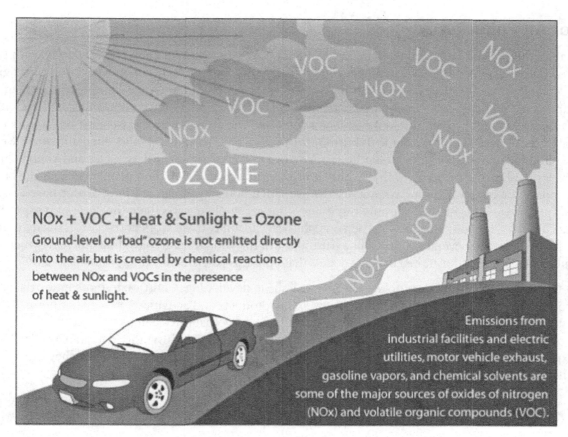

Figure 7.1. *Tropospheric ozone forms as a secondary pollutant when nitrogen oxides (NOx) and volatile organic compounds (VOCs) react in the presence of sunlight.*[1]

with respiratory conditions the greatest. During high ozone days, urban areas often experience an increase in emergency room visits for acute asthma attacks in school-age children.

There are three main factors that determine the level of **tropospheric ozone** (also called ground-level ozone) that is present in our atmosphere: *temperature*, *emissions*, and *population*. When the atmosphere temperature rises, sunlight exposure increases the amount of chemical reactions that result in ozone production. When both emissions and population size increase in a particular area, so will the presence of ozone, which is why most places with the highest levels of ozone are found in urban areas like Los Angeles, Houston, or Dallas. Therefore, the highest levels of tropospheric ozone are typically seen during warm summer months in urban areas when there is little wind or cloud cover.

Another pollutant is **particulate matter (PM)**, also called particle pollution, which is a mixture of extremely small particles and liquid droplets suspended in the air. Under the NAAQS, states are required to monitor fine particles between 10 and 2.5 micrometers (**PM$_{10}$**) and particles smaller than 2.5 micrometers (**PM$_{2.5}$**). Keep in mind that it takes 1,000,000 micrometers to make one meter, so these particles are extremely small! PM$_{10}$ particles are small enough to pass in into the respiratory system without being removed by coughing or sneezing.[2] PM$_{2.5}$ particle are small enough to be able make its way deep into respiratory systems, causing respiratory difficulties. Like ozone, temperature, emissions, and population all play a role in the levels of particulate matter found in the air we breathe.

---

[1] AirNow. 2014. "Ozone." AirNow.gov. Accessed January 10, 2014. http://www.airnow.gov/index.cfm?action=aqibasics.ozone
[2] U.S. Army Center for Health Promotion and Preventative Medicine. 2002. "Particulate Matter." Deployment Health Library.

Indoor Air Pollution

Indoor air pollution is an issue mainly in the developing world, where indoor cookstoves, coupled with poorly ventilated homes, can result in elevated concentrations of particle pollution, sometimes at 100 times greater than acceptable levels. In these homes, the fuel sources most often used are the cheapest: wood, dung, crop waste, and coal. While these sources may be inexpensive, they do not combust completely, and they release a large amount of partially combusted particulate matter into the air. Because these cookstoves are also often the sole source of heat, the stoves are kept burning 24 hours a day. The groups most often at risk are women, who are often the ones responsible for tending to the stove and cooking, as well as babies and young children with developing lungs. According to the World Health Organization, 1.6 million people per year die because of poor indoor air quality. Chronic respiratory disease, pneumonia, asthma, and lung cancer have also been linked to poor indoor air quality. Chronic respiratory illness can often perpetuate the cycle of poverty in developing countries, because it decreases a person's physical capability to do work and secure a steady source of income for their family. Several countries like Australia and the United States have implemented programs to swap out old woodstoves for more efficient models that emit less particulate matter, or none at all.[3,4]

However, poor indoor air quality is not exclusively an issue in developing countries. In more industrialized nations, poor indoor air quality is caused by many everyday products and activities you may not normally think about, such as the spraying of chemicals and perfumes, the use of laundry detergent, and burning candles. Many potentially damaging indoor air pollutants tend to be associated with synthetic fragrances. While these small exposures generally will not result in acute toxicity (i.e., asthma attacks, even death), they can accumulate within our bodies as chronic exposure continues over years or decades.

ATMOSPHERIC DEPOSITION OF POLLUTANTS

While NAAQS monitors *some* outdoor air pollutants, those six pollutants are not the only ones making it into our air. The pollutants may not pose a threat to our respiratory health, but they could later become a health concern if they go through additional chemical reactions that may turn them into more harmful compounds.

Atmospheric deposition is the transfer of particles suspended in the air onto land or water. This can happen through *dry deposition*, in which particulate matter settles out of the atmosphere directly onto the surface, or through *wet deposition*, when particulate matter mixes with moisture in lower atmosphere and reaches the surface with rain. Due to atmospheric wind movement patterns, particulate matter does not always deposit in the same place where it was originally introduced into the atmosphere. In fact, scientists are starting to take note of elevated mercury levels in the Pacific Ocean, most likely due to large coal-fired power plant production in China.[5]

TROPHIC LEVELS

You may have heard that elevated mercury levels are often associated with fish and seafood. This is why it is recommended that we limit our consumption of some fish species, especially bigger species, such as tuna, to no more than once a week. Why are bigger fish more susceptible to mercury accumulation? And why is it important that humans limit their seafood intake?

---

[3] NSW. 2011. "Air Pollution: Particulate Matter (PM$_{2.5}$)." New South Wales Government Office of Environment and Heritage. Sydney, Australia. http://www.epa.nsw.gov.au/resources/air/110456Airpollution.pdf

[4] EPA. 2013. "Agencies – Changeout Guide." United States Environmental Protection Agency. http://www.epa.gov/burnwise/how-to-guide.html

[5] Long Wanga, et al. 2014. "Source appointment of atmospheric mercury pollution in China using the GEOS-Chem model." *Environmental Pollution* 190, 166–175. doi:10.1016/j.envpol.2014.03.011

Recall the concept of food chains from high school biology. A food chain describes the order of predation events within an ecosystem. For example, we know that a rabbit will eat plants, and that a fox will eat the rabbit. Each step in the food chain represents a different trophic level. A **trophic level** describes the position an organism occupies in a food chain, especially with respect to energy transfer. The lowest trophic level consists of organisms that produce their own energy (producers), such as plants and phytoplankton. Organisms down the food chain from plants are considered consumers, because they have to eat other organisms in order to gain energy.

The trophic pyramid consists of producers at the base, followed by primary consumers (herbivores), secondary consumers (carnivores), tertiary consumers (higher carnivores), and so forth. However, only about 10% of the energy transferred from one trophic level to the next is converted into biomass. This means that *each organism in the food chain must eat ten times the amount of food per unit of body mass relative to the organism directly below it on the food chain.* Due to this inefficiency, most ecosystems will not have more than 5 or 6 trophic levels.

BIOACCUMULATION AND BIOMAGNIFICATION

Organisms that live at any particular trophic level accumulate (**bioaccumulate**) the toxic chemicals in their tissue, because their bodies don't know what to do with these compounds. Bioaccumulation occurs within one individual by one of two ways:

1. If the body mistakes the toxin for an essential nutrient, then the toxin will bioaccumulate. A historical example of this process took place after the Chernobyl nuclear power plant incident in 1986, which resulted in the exposure to radioactive strontium over a large portion of the area. Strontium is structurally similar to calcium, so cows' bodies in the vicinity mistook the toxin for calcium and incorporated the compound into their milk. As a result, humans and calves were receiving high dosages of strontium through "radioactive" milk.

2. If the toxin is not mistaken for an essential nutrient and is not ejected by the individual, then it will be become "trapped" in the fatty tissues or bound to certain proteins. This happens with several heavy metals, including mercury, chromium, and lead. Toxic chemicals are either water or oil soluble. Oil soluble toxins move into fatty tissues, while water soluble toxins move more readily throughout the body, since water is found nearly everywhere in living tissue.

When an organism in a higher trophic level consumes an organism in a lower trophic level, it is also consuming the toxic chemicals in the tissue. While consumers only retain 10% of the energy they gain from the food they eat, they retain almost *all* of the toxins accumulated in the tissues of their food. Thus, if a pelican has to consume 10 pounds of fish for every pound of its own body mass, and if each pound of fish is contaminated with toxins, the pelican will accumulate 10 times the concentration of toxins as each individual fish had. The increase in the concentration of toxins upward through trophic levels is called **biomagnification**.

MERCURY

Mercury is a carcinogen that bioaccumulates in the fatty tissue of organisms and biomagnifies upward through trophic levels. Mercury poisoning can cause birth defects in unborn children and neurological damage in adults. The largest anthropogenic source of mercury in the environment is from coal-fired electricity plants (Figure 7.2). U.S. power plants collectively emit approximately 48 tons of inorganic mercury into the air each year.[6] Atmospheric deposition sends the mercury into our reservoirs, where bacteria metabolize it and convert it to organic mercury. While the inorganic mercury that leaves smokestacks is not harmful to humans, organic mercury can be extremely toxic at high concentrations. Mercury in marine environment is biomagnified in the food chain, resulting in high mercury concentrations in top predators such as tuna.

---

[6] Bowen, Brian H. and Marty W. Irwin. 2007. "Basic Mercury Data and Coal-Fired Powerplants." Indiana Center for Coal Technology Research, Purdue University.

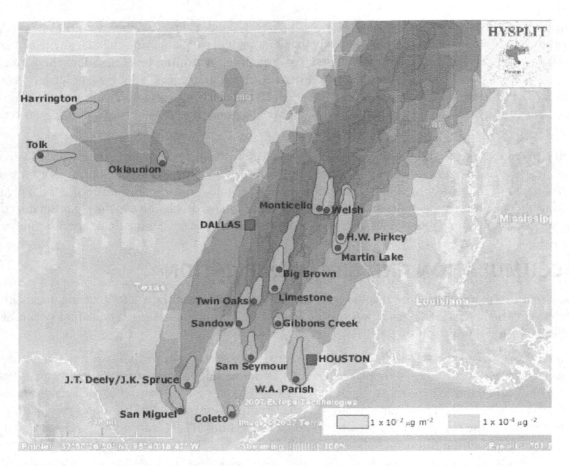

Figure 7.2. *Location of coal-fired power plants in Texas with pollution plumes on November 5, 2006.*

Name:

LAB 7 WRITE-UP: AIR POLLUTION AND BIOACCUMULATION

In Exercises 1 and 2, you will use a computer model to manipulate the formation of ground-level (i.e., tropospheric) ozone and particulate matter.

Exercise 1: Air Pollution and COVID-19 (3 points)

The COVID-19 pandemic has caused a lot of lifestyle changes around the word due to quarantine efforts to hinder disease spread. Due to these changes, human activity has decreased and resulted in improved air quality in certain areas. Write a 500-word essay selecting two of the air contaminants discussed in lab today and how their concentrations have changed as a result. You are required to use at least three sources (news, academic, etc.) to back of your claims and may use the aid of figures (ie. before/after photos, pollution maps).

Exercise 2: Air Pollution in Your Zip Code (3 points)

1. Go to https://www.airnow.gov/aqi/aqi-basics. What is the AQI and how does it work? **(0.5 point)**

2. What are the five major pollutants used to determine AQI? **(0.5 point)**

3. Now go to https://www.airnow.gov/ and enter your zip code.
    What is the current AQI for your zip code? What is the Primary Pollutant(s) for your area? What warnings are given for each pollutant? **(0.5 point)**

4. What is the forecast pollutant for the next few days? Are there any new pollutants or new warnings present? **(0.5 point)**

5. Ground-level ozone is a common air pollutant in urban areas. What is the difference between ground-level ozone and stratospheric ozone? How is ground-level ozone formed? More information about ozone can be found at https://www.epa.gov/ground-level-ozone-pollution/ground-level-ozone-basics#effects. **(1 point)**

Exercise 3: Mercury and Other Fish Advisories (4 points)

1.   Go to the following link https://www.epa.gov/choose-fish-and-shellfish-wisely/fish-and-shellfish-advisories-and-safe-eating-guidelines to learn more about fish consumption advisories.

   a.   What are fish advisories? **(0.5 point)**

2.   Now go to the following link http://www.dshs.texas.gov/seafood/advisories-bans.aspx to find the most current fish consumption advisories in the state of Texas or http://fishadvisoryonline.epa.gov/General.aspx for fish advisories in other states.

   Choose a body of water (possibly from your hometown) and download the advisory links to find out what contamination was present in the fish, as well as whether consumption is permissible or limited for certain species.

   a.   Which body of water did you choose? **(0.5 point)**

   b.   What contaminants are present? **(0.5 point)**

   c.   What fish are under these advisories? **(0.5 point)**

3.   Once you find out what these contaminants are, do some extra research on the internet to find out the side effects and concerns associated with these compounds (page 11 http://fishadvisoryonline.epa.gov/docs/NLFA_Advisories_Where_You_ Live_User_Guide.pdf and http://www.atsdr.cdc.gov/ has some basic information you may find helpful).

   a.   Do any of these compounds bioaccumulate or biomagnify? What type of effects do they have on wildlife and humans? How does the presence of these compounds make you feel about the body of water and fish? **(2 points)**

$\mathcal{L}ab$ VIII

Soil and Water (Lab Day)

SOILS

Before we can understand water, we must first explore the soil characteristics that control water as it moves through the environment. Soils are the result of local geology, regional climate, topography, and the organisms that live (or once lived) in close relationship within that soil. Understanding soils opens a window through which we can understand geologic and even human history. First, let's familiarize you with the properties of soils that govern how water moves across the landscape.

Soils and Their Interaction with Water

Generally defined, the term **soil** refers to unconsolidated, loose covering of fine rock particles that differ from their parent material. Soils are comprised of a mixture of organic and inorganic (mineral) materials. To a large extent, the soils within a watershed will define local hydrology (the movement of water across a landscape). **Watershed** is the term used to describe the geographic area of land that drains water into a particular river, stream, or body of water. The highest ground around the watershed forms its boundaries.

There are several paths that water can take to reach the river and even more ways in which the water can be intercepted, evaporated, or transpired (in other words, lost to the atmosphere) before it reaches the river (Figure 8.1). What we will be considering now is how different soil types and land uses affect water on its path toward a river.

We will start from the moment a single raindrop hits the soil. One of two things can happen to this single drop: it can (1) *infiltrate* into the soil or (2) *run off* downslope. During **infiltration**, water percolates through

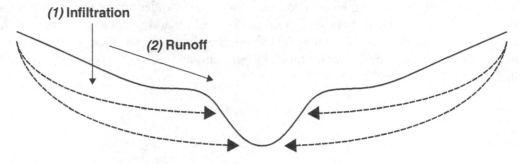

Figure 8.1. *Rainfall can either infiltrate or run off toward the watershed low point.*

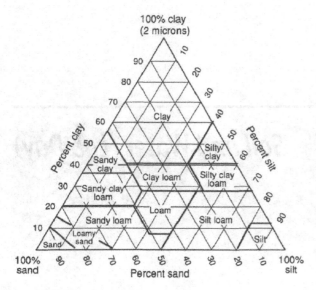

Figure 8.2. *The soil texture triangle.*
Source: USDA NRCS

soil particles until it either enters the groundwater or flows downslope in the soil matrix. In both cases, the water moves very slowly through a medium toward a low point such as a river. **Runoff** water travels more quickly toward a topographic low point or a body of water. Whether the water infiltrates or runs off down slope depends on three main environmental characteristics: soil texture, vegetation, and land use.

Soil Classification Based on Soil Texture

Soils are porous mixtures of mineral particles, organic matter, air, and water. The physical properties of a given soil are dependent on the nature and amount of each constituent. Mineral particles determine a soil texture and typically make up about one-half the volume of a soil in good condition for plant growth. Mineral particle sizes can be divided into three groups called **soil separates**. The three soil separates are *sand* (2.0–0.05 mm), *silt* (0.05–0.002 mm), and *clay* (less than 0.002 mm). **Soil texture** is determined by the relative proportions of sand, silt, and clay in a soil's mineral composition.

The triangle diagram in Figure 8.2 allows you to use information about proportions of soil separates to determine the appropriate name for the soil texture. For example, a soil made up of 40% sand, 30% clay, and 30% silt is a clay loam.

Sandier soils are termed *coarse-grained* soils because sand is a large, coarse particle. *Fine-grained* soils have higher proportions of clay in them because clay is a very small particle. Soil texture is one of the most important descriptors of a soil because it determines many other soil properties. Texture will affect a soil's agricultural potential, the natural vegetation growing within it, nutrient storage potential, and other physical, chemical, and biological properties. Presently, we are interested in how soil texture affects the movement of water.

Knowing the texture of a soil allows us to predict how water will move *into* (*infiltration*) or *over* (*runoff*) any particular soil. Coarse soils, such as sandy loams or loamy sands, will have high infiltration rates and low runoff, while fine-grained soils, such as clay and clay loams, will have low infiltration rates and high runoff. Let's formalize this concept by organizing different soil textures into a classification system determined by a soil's infiltration rates.

Hydrologic Classification of Soils

Soils are classified into four *hydrologic classifications*: A, B, C, and D. "A" group soils have the highest infiltration rates, while "D" group soils have the lowest infiltration rates. For example, a sandy soil might be in the "A" group, while a clayey soil would be in the "D" group. Table 8.1 summarizes the four hydrologic classifications.

It is important to note that the existing level of soil saturation will also determine the amount of water that will infiltrate or run off. Saturated soils have less room (pore space) for additional water to occupy, while drier soils have more room. Although clays have a low infiltration rate, the small pore spaces between clay particles cause clays to retain the water that infiltrates for much longer than sandy soils. This causes clayey soils to swell when saturated and to shrink when dried.

Table 8.1. *The four hydrologic classifications for soils. The classifications represent a spectrum where "A" group soils have the lowest runoff potential and the highest infiltration and "D" group soils have the highest runoff potential and the lowest infiltration.*

Hydrologic Classification Chart

A	Limited runoff; high infiltration even when thoroughly wetted (>0.30 inches/hour). Usually deep, well-drained sands and gravels.
B	Moderate infiltration capacity when thoroughly wetted (0.15–0.30 inches/hour). Moderately deep, well-drained, and generally fine to moderately coarse grained (e.g., sandy loam).
C	Low infiltration capacity when thoroughly wetted (0.05–0.15 inches/hour). Generally fine-grained soils (loamy clays) or with an impeding layer.
D	Very low infiltration; high runoff. Clay soils with high swelling potential.

The Effect of Vegetation and Land Use on Infiltration Rates

Vegetation increases infiltration and decreases runoff in three general ways:

1. Plant roots open up pores in the soil, increasing the number and size of infiltration pathways.

2. Leaves, brush, and low-lying plants slow down surface runoff and allow more time for runoff to infiltrate.

3. The aboveground vegetation protects the soil from the force of falling raindrops, which compact soils slowly by continuously and decreases infiltration potential.

Humans—and the variety of ways in which they use soils—tend to *decrease* infiltration rates and increase runoff. Human activities, especially agriculture, compact soils with heavy machinery (tractors, cars, and so on) or cow hooves. Poor soil management techniques, such as leaving soil fallow with no vegetation, increases runoff. Humans construct impermeable surfaces such as concrete parking lots and rooftops, preventing water from infiltrating into soil. Water running off impermeable surfaces also experience increased runoff velocity, which increases the potential for erosion when it finally makes it onto the soil.

Infiltration, Runoff, and Surface Water

Runoff water moves downslope very rapidly, picking up debris, pollutants, and eroded soil particles on its journey to surface water bodies, such as streams. After the onset of a large storm, the soils will absorb (through infiltration) as much water as they can hold, and the remaining water will run down slope (Figure 8.3).

After the upper layer of the soil reaches its saturation point, there will be a discharge of muddy water as it runs off toward streams. If the soils are fine-grained, the volume of runoff will be very high and the stream will then swell, often causing flash floods. This scenario is depicted by the *hydrograph* in Figure 8.4a. Notice

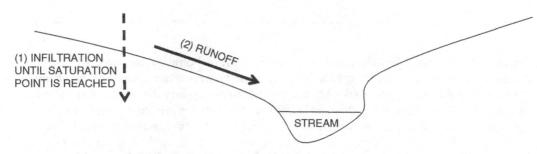

Figure 8.3. *Rainfall will infiltrate into the soil until the upper layer of the soil is saturated and cannot hold any more water. The remaining rainfall will runoff down-gradient.*

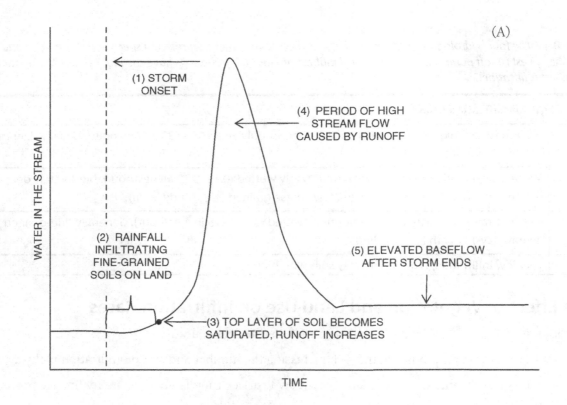

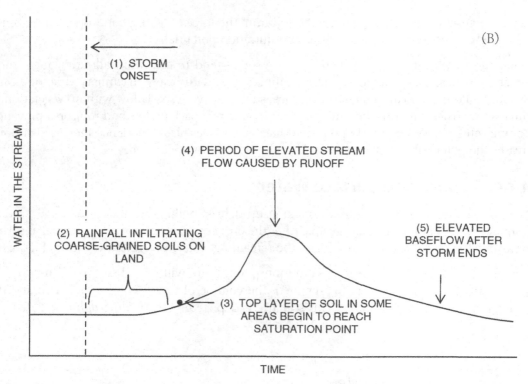

Figure 8.4. *(A) A hydrograph of a stream in a dominantly fine-grained soil watershed. Infiltrated water will quickly saturate the top layer of soil, and the remaining water will run off into the stream, which will experience flash flooding. After the storm ends, the stream level will soon return to baseflow. The new baseflow will be slightly elevated from pre-storm conditions due to throughflow (horizontal percolation of water in the subsurface soils toward a low point) and increased contribution from the recently recharged groundwater. (B) The hydrograph for a stream in a coarse-grained dominated watershed shows a longer period of infiltration and a less dramatic flooding in the stream.*

that moments after the rainfall event, the amount of water flowing through the stream increases suddenly and dramatically. This is characteristic of fine-grained soils.

In contrast, coarse-grained soils have higher infiltration rates and take longer for the top layer of soil to become saturated. Furthermore, when coarse-grained soils do become saturated, water drains out of the top layer much faster than in fine-grained soils, opening up pore space for more rainfall to infiltrate. This means that the hydrograph for coarse-grained soils (Figure 8.4b) shows much less dramatic flood stages for the stream.

Watersheds containing soils with high infiltration rates are characterized by *perennial streams*. Perennial streams flow year-round and are supplied by groundwater discharge. In contrast, watersheds with high runoff rates are characterized by *intermittent streams* and *ephemeral streams*. Intermittent streams typically only contain water during wet seasons, whereas ephemeral streams usually only have a flow during the flash flooding period immediately after rainfall events.

Water flow in urban watersheds has severely been altered by intense development of impervious covers (e.g., roads, parking lots, and buildings). Prior to development, natural landscape features (e.g., vegetative cover and soil type) in these watersheds allowed for regulated flow of water either through subsurface or overland flow. However, with the development of virtually impermeable surfaces, storm water is now routed through storm water systems into streams and rivers at high volumes and velocities. Note that this flow only occurs around storm events, which simulates similar characteristics to fine-grained soils as depicted in Figure 8.4a. Therefore, stream flow is highly variable, affecting ecosystems downstream. Moreover, the development of impervious covers allows for the accumulation of pollutants on these surfaces. Thus, a flux of contaminants enters waterways during and just after a storm event, degrading the quality of our water resources.

WATER QUALITY

Now that you understand the role of soil in the hydrologic cycle, let's look at the *quality* of our water. Water is essential for life and plays a vital role in the proper functioning of the earth's ecosystems. The pollution of water has a serious impact on all living creatures and can negatively affect the use of water for drinking, household needs, recreation, fishing, transportation, and commerce. As the human population grows, more and more water is appropriated to serve human needs (at the expense of nonhuman entities, including organisms and ecosystems).

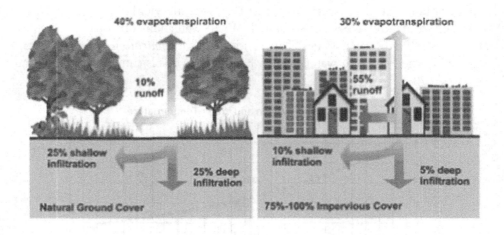

Figure 8.5. *Comparison of water flow in natural vs. urbanized landscapes.*
Source: Adapted from EPA, 2003.[1]

---

[1] United States Environmental Protection Agency. 2003. "Protecting Water Quality from Urban Runoff." *Urban Nonpoint Source Fact Sheet.*

Dissolved Oxygen

Aerobic organisms including humans, animals, and plants require oxygen to survive. Terrestrial organisms live in an atmosphere with the relatively constant level of gaseous oxygen (about 21%). Aquatic organisms also need oxygen dissolved in water. However, sufficient oxygen concentrations in an aquatic environment are not to be taken for granted, as dissolved oxygen (DO) fluctuates in seasonal and daily cycles. Several factors can influence the level of DO in a water body.

1. The *temperature* of the water can affect DO. As water temperature increases, the capacity of water to hold DO decreases (Figure 8.5).

2. *Photosynthesizing plants* will increase DO. During the day, plants replenish the waters with oxygen as a by-product of photosynthesis, while at night, this process stops, and oxygen levels decrease.

3. *Turbidity*, or cloudiness of the water, can decrease photosynthetic rates during the day and thus decrease DO additions.

4. *Riffles*, defined as areas of a distinct change in gradient where flowing water can be observed, and rapids will add DO to waters through the physical mixing action (similar to what is going on when you whip egg whites).

The worst conditions for a fish might be a hot (high temperature), stagnant (no riffles), murky pool (high turbidity) with no plant life.

Turbidity

Turbidity is the measurement of the cloudiness of water, and indicates the amount of suspended solids in the water column. In addition to affected DO content, turbidity is an indicator of *the* quality of water as turbid water can contain contaminants that could cause illness. In this lab, turbidity will be measured in nephelometric turbidity units (NTU) using a turbidity meter. NTUs are a measure of the amount of scattering of light that is caused by the suspended solids. In the United States, drinking water should be less than 1.0 NTU at the drinking water treatment plant outlet.

Nitrate

Nitrogen is an essential nutrient for both plant and animal life. In fact, a main component in fertilizers is nitrogen, in the form of nitrate. Fertilizer is used to increase growth and yield of crops. Unfortunately, fertilizer does not stay contained in the area where it is applied. When water falls on soils that have been fertilized, these nitrates can be washed into water bodies. This can lead to *eutrophication* (refer to *Lab 3: Food*), which decreases the dissolved oxygen levels in water bodies. High nitrate levels can also impact human health and

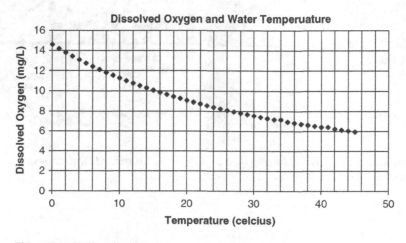

Figure 8.6. *Dissolved oxygen as a function of water temperature.*

can be fatal, especially in young children. Under the Safe Drinking Water Act in 1974, the EPA was given the right to monitor and set drinking water standards for harmful contaminants. Under the Safe Drinking Water Act, nitrate levels of 10 ppm or under are not harmful to human health.

Phosphate

Phosphates, which are also an essential nutrient, are naturally occurring in rocks and are also used artificially in plant fertilizers. Phosphates make their way into waterways when water runs off over phosphorus-rich rocks or through the application of synthetic fertilizers. In natural freshwater bodies, phosphate levels are about 0.02 ppm. In water bodies that are influenced by anthropogenic sources, phosphate levels can be much higher. Common sources for anthropogenic phosphorus are from water outfalls (e.g., of untreated sewage), fertilizers, and detergents.

While plants and animals are dependent upon phosphorus as an essential nutrient, like nitrogen, high levels of phosphorus can lead to eutrophication and declining water quality.

Total Alkalinity

Alkalinity is a measure of a body of water's ability to neutralize acid. Bicarbonates and carbonates in water have the ability to remove hydrogen ions (H^+) and lower the acidity of the water. Water bodies that lack the ability to neutralize acid will experience an immediate change in pH if acidifying substances are introduced. This is important especially in instances of exposure to acid rain or wastewater discharge.

pH

pH is the measurement of how acidic (low pH) or how alkaline (also known as "basic," high pH) a substance is. pH is based on a negative log scale that measures the concentrations of H^+ in a solution. Equation 8.1 shows the mathematical equation for determining pH.

$$pH = -\log[H^+] \tag{8.1}$$

A pH scale is shown in Figure 8.7. Notice the scale ranges from 0 to 14, and *pure water has a pH of 7.* As water becomes more acidic, it is less able to support life. Conversely, water that is too alkaline can be just as inhospitable as acidic water.

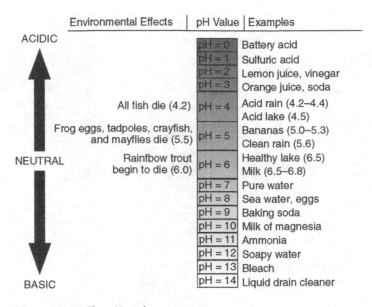

Figure 8.7. *The pH scale.*

Look at the scale in Figure 8.7. A "healthy" lake is *slightly* acidic with a pH of 6.5. At a pH of 6.0, trout populations in the lake will begin to experience adverse impacts. This is because pH is a log scale, meaning that the shift from 7.0 to 6.0 means represents a difference of an order of magnitude in the hydrogen ion concentration. "Natural" rainfall is also *slightly* acidic, and most of our food crops prefer a *slightly* acidic soil. However, increasing the acidity too much (acid rain) will kill vegetation.

Name:

LAB 8 WRITE-UP: SOIL AND WATER IN LAB

This is the first of a two-part lab. Today you will work in a lab environment to practice basic techniques for testing soil and water quality. During the second part of this two-part lab, you will be putting these tools to use at a field site.

Exercise 1: Soil Texture Analysis

You will be presented with three unknown soil samples and some water vials with a dropper. This exercise will give you the opportunity to master the art of identifying unknown soil types.

1. **Preliminary Observations:** First we will analyze the ability of the soil to hold water. To do this, take a moderate amount of your soil sample in hand, and add a small amount of water (a drop) and mix. Continue to do this until you can form a ball with these samples. How many drops of water were required to form a ball of your unknown sample? Based on these preliminary observations, do you think your soil sample would be clay dominated or sand dominated? Record your answers on the data sheet provided.

2. **Ribbon Testing:** Now that we have saturated the soil enough to manipulate it, we can perform a ribbon test. A ribbon test involves pressing the balled-up sample between your thumb and index finger while gently pushing upward. If the soil is clay dominated, the sample should form a "ribbon" of soil. Using the ribbon test flow chart, determine the type of soil you have and record it on your worksheet.

3. **Soil Hydrologic Analysis**: Based on the results of your ribbon test, which hydrologic group (A through D) do you think best characterizes your sample? Record your results on your data sheet.

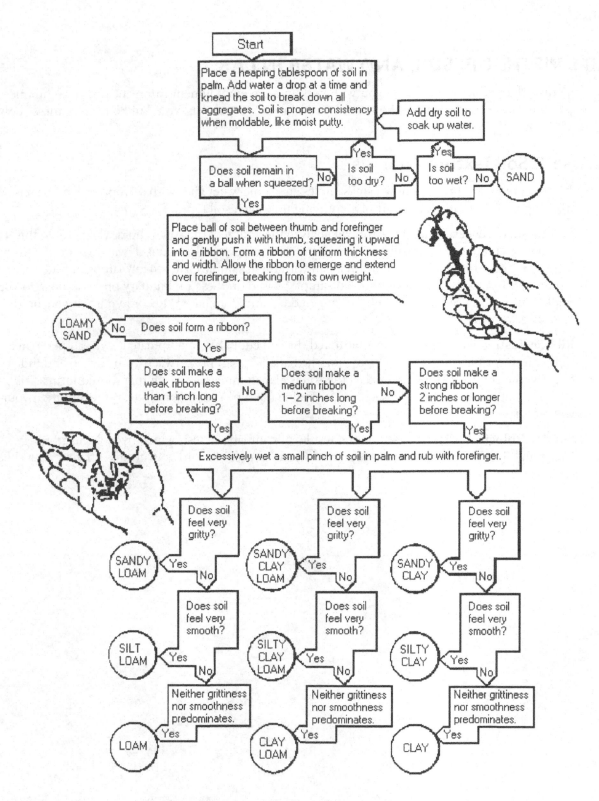

Start

Place a heaping tablespoon of soil in palm. Add water a drop at a time and knead the soil to break down all aggregates. Soil is proper consistency when moldable, like moist putty.

Add dry soil to soak up water.

Does soil remain in a ball when squeezed? — No → Is soil too dry? — Yes ↑ / No → Is soil too wet? — Yes ↑ / No → **SAND**

Yes ↓

Place ball of soil between thumb and forefinger and gently push it with thumb, squeezing it upward into a ribbon. Form a ribbon of uniform thickness and width. Allow the ribbon to emerge and extend over forefinger, breaking from its own weight.

LOAMY SAND ← No — Does soil form a ribbon?

Yes ↓

Does soil make a weak ribbon less than 1 inch long before breaking? — No → Does soil make a medium ribbon 1–2 inches long before breaking? — No → Does soil make a strong ribbon 2 inches or longer before breaking?

Yes / Yes / Yes

Excessively wet a small pinch of soil in palm and rub with forefinger.

Does soil feel very gritty? | Does soil feel very gritty? | Does soil feel very gritty?

SANDY LOAM ← Yes / No ↓ | **SANDY CLAY LOAM** ← Yes / No ↓ | **SANDY CLAY** ← Yes / No ↓

Does soil feel very smooth? | Does soil feel very smooth? | Does soil feel very smooth?

SILT LOAM ← Yes / No ↓ | **SILTY CLAY LOAM** ← Yes / No ↓ | **SILTY CLAY** ← Yes / No ↓

Neither grittiness nor smoothness predominates. | Neither grittiness nor smoothness predominates. | Neither grittiness nor smoothness predominates.

LOAM ← Yes | **CLAY LOAM** ← Yes | **CLAY** ← Yes

Modified from: Thien, Steven J., Kansas state University, 1979 Jour. Agronomy education.

Exercise 2: Water Quality

In this exercise, we will conduct several tests on your classroom tap water. Tap water is not pure water, as it contains several minerals and chemicals such as chlorine and fluoride. Depending on the source of your tap water, it may also contain some degree of visible sediment.

1. Fill a plastic cup with water from the sink in the back of the room

2. Dip strip directly into cup and swirl 2 times

3. Remove strip with pads face up (DO NOT SHAKE OFF EXCESS WATER)

4. Compare to color chart on the back of the test strip bottle

5. Immediately read General Hardness (GH) and Carbonate Hardness (KH)

6. Wait 30 seconds and read pH, Nitrates, and Nitrites

7. Record your results on your data sheet.

Make sure that you understand everything done in this lab. This is your "practice run". Next week, you will be out in the field performing the same analysis for the soils and surface water at your field site and collecting the data for a formal lab report.

Name:

LAB 8: SOIL AND WATER DATA SHEET

Exercise 1: Soil Texture Analysis

Sample	Soil Type	Hydrologic Group	**Description** (include number of water drops required to form a ball, ribbon test length results, color, and anything else you found to be unique)
A			
B			
C			

1. Using your lab manual and lecture notes, briefly explain how soil texture and hydrologic group are related. Which groups are clay dominated? Which groups are sand dominated? Explain the relationship between particle size, porosity, and infiltration rates. (**1 point**)

_____

_____

_____

_____

_____

_____

2. How do humans alter the infiltration/runoff rates of a landscape? (**1 point**)

_____
_____
_____
_____
_____
_____

Exercise 2: Water Quality

Record your water quality results in the table below. (**4 points**)

General Hardness (GH)	
Carbonate Hardness (KH)	
Nitrates (NO_3^-)	
Nitrites (NO_2^-)	
pH	
Phosphates (PO_4^-)	**4 ppm**

Lab IX

Soil and Water (Field Day)

Introduction

Fieldwork is an integral part of being an environmental scientist. It is often dirty, too hot or too cold, fraught with unforeseen challenges, and thrilling, all at the same time. Since it can often be a challenge to collect quality data in a limited amount of time, time in the field must be maximized to its fullest. Preparation for fieldwork is vital, which is why we spent Lab 8 developing the necessary understanding and techniques to analyze soil and water. Today you will apply the techniques you practiced in the laboratory in a field setting. Your instructor will take you on a field trip to a stream area, where you will collect soil and water data for analysis.

Name:

LAB 9: SOIL AND WATER ANALYSIS IN THE FIELD DATASHEET

Part I: Making Observations

A. Before you begin your fieldwork, be sure to take down the appropriate metadata. **Metadata** gives information about other data. This includes the date, time, location, and even weather conditions at the time of data collection. The absence of metadata may make the data you collect in the field meaningless, especially if you are working on a long-term research project. Record your metadata in the space below. **(0.5 point)**

Part II: Making A Cross-section of Your Portion of the Stream

A. Using a tape measure, record the following characteristics of your stream's cross-section (the area between upland environments) in centimeters or meters. Refer to the example drawing in Figure 9.1. **(0.5 point)**

Width of cross section: _____

(*From highest point to highest point*)

Width of stream: _____

Averaged depth of the stream: _____

(*Averaged over 3 different points*)

Depth measurement #1 _____

Depth measurement #2 _____

Depth measurement #3 _____

Height of left streambank _____

Height of right streambank _____

B.   Sketch the slopes going down to the stream. On this sketch, note the placement of vegetation, concrete, gravel, etc. at your cross-section. Include the total width of the cross-section (from highest point to opposite highest point), the height of the slopes above the stream, and the width and depth of the stream (measure at multiple spots if the depth is variable). Be sure to include units. **(2 points)**

C.   Record your observations about the flow in stream. Were there any riffles? How fast was the water moving? Which direction was the water moving in? **(0.5 point)**

Part III. Soil Texture Analysis

A.   Identify your two sample locations on your stream cross-section drawing.

B.   Record observations about your soil sample in the space below. Was the soil wet or dry prior to examination? Was the soil texture the same at various places along your portion of the stream? What was the soil color? Was there a lot of organic matter (i.e. leaf litter, twigs) in your soil samples? **(0.5 point)**

C.   Each person in your group should collect a small amount of the soil and determine the soil texture for both sides of the stream bank using the flowchart in Lab 8.

1.   What is the texture of your soils? Left bank:_____, Right bank:_____ **(0.25 point)**

2.   In what hydrologic class would you place the soils? Left banks:_____, Right bank:_____ **(0.25 point)**

Part IV: Water Quality

A. Using the techniques you practiced in Lab 8, collect the following data from the water in your stream: **(0.5 point)**

	Stream	Classroom/tap water
General Hardness (GH)		
Carbonate Hardness (KH)		
pH		
Nitrate (ppm)		
Nitrite (ppm)		
Phosphates (PO_4^-)		
Turbidity (cloudiness)		

B. Aquatic life observations. What do you see in the water at your stream? Are there any aquatic plants? Algae? Are there invertebrates, like snails, or vertebrates, like fish or frogs? **(0.5 point)**

C. Where is this water coming from (e.g. storm drains, residential runoff, natural spring)? What are the immediate upstream and downstream uses of the stream (that you can see from our stream location)? How does this section of the stream differ from the surrounding sections? **(0.5 point)**

Part V. Discussion Questions

1. What observations or educated guesses can you make about the infiltration rate and runoff of the soil at your portion of the stream? How do you think the nearby conditions (e.g. soil type, vegetation, concrete, steep or flat slopes, urban influences) affect infiltration and runoff in the stream? **(1 point)**

2.   What impact do the conditions outlined above have on the stream's water quality? How does the stream's water quality compare to the lab's water quality? (**1 point**)

3.   Would you assume the dissolved oxygen (DO) content of the tap water or stream water to be higher? What about the DO content of the stream water compared to the channelized stream portion upstream of our field location? Explain the reasoning behind your estimates based on the stream conditions. (**1 point**)

4.   How would you expect the observed stream to differ from a stream in a nature area in terms of infiltration/runoff rates and water quality? (**1 point**)

$\mathcal{L}ab$ X

The Flint Water Crisis

Introduction

Since the beginnings of civilization, human society has been dependent upon access to water. Maintaining access to clean water with adequate sanitation and providing disposal systems for sewage become more pressing issues as populations grow. Advanced technologies have been used throughout history to respond to these concerns. The Etruscans, who lived in ancient Italy, built the first sewers in 50 BC, and in 80 AD, Roman senators voted to protect water sources during summer months so that streets and sewers could be flushed.

BACKGROUND

The Flint Water Crisis

For more than a century, the Flint River, which flows through the heart of the city of Flint, Michigan, served as an unofficial waste disposal site for many local industries, car factories to meatpacking plants. The river also received raw sewage from Flint's waste treatment plant, agricultural and urban runoff, and toxics from a number of landfills.

By the mid-20th century, Flint – the birthplace of General Motors – was a booming city of almost 200,000 people, many employed by the automobile industry. The landscape had changed dramatically by the 1980s. Rising oil prices and a decline in the auto industry meant an exodus of workers. Flint went into a prolonged period of decline.

By 2011, Flint was strapped for cash and burdened with a $25 million deficit. The Governor appointed an emergency manager to oversee and cut city costs. The decision was made to end the city's five-decade practice of piping treated water for its residents from Detroit in favor of a cheaper alternative: temporarily pumping water from the Flint River until a new water pipeline from Lake Huron was built. Although the river water was highly corrosive, Flint officials failed to treat it, and lead leached out from aging pipes into thousands of homes.

Name:

LAB 10 WRITE-UP: "NOVA: POISONED WATER"

Watch "NOVA: Poisoned Water" (PBS, 2017) and answer the following questions as you watch the documentary. The documentary can be found on Netflix and at https://www.pbs.org/video/poisoned-water-jhhegn/ (If the website indicates you need to pay switch internet browsers).

1.   Why did Flint, Michigan, switch water sources? **(0.5 points)**

2.   How do treatment plants remove particles from river water? How do they remove chemicals from river water? **(0.5 points)**

3.   Why did GM switch back to Lake Michigan water? **(0.5 points)**

4.   How much lead was found in Susan's house? How much is allowed by the EPA's Lead and Copper rule? **(0.5 points)**

5.   Why is lead toxic to people? **(0.5 points)**

6.   What are the four ways lead can get into drinking water? **(0.5 points)**

7.   What is scale? Why is it important? **(0.5 points)**

8. What is corrosion control treatment? What was Flint using for anti-corrosion? **(0.5 points)**

9. How much lead was found in the Walter's home after 20 liters of water had been flushed through their sink? What level of lead is considered hazardous waste? **(0.5 points)**

10. Why did Washington DC have elevated levels of lead? **(0.5 points)**

11. How many children were potentially exposed to lead in the water in Washington DC? **(0.5 points)**

12. What were the results of Virginia Tech's study of Flint's water? What item was recommended for the citizen to buy? **(0.5 points)**

13. What were citizens told to do before sampling their water? **(0.5 points)**

14. What did the study at Flint's Hurley Children's Hospital discover? **(0.5 points)**

15. What is Legionnaires' disease? **(0.5 points)**

16.   How did the Legionnaires' disease outbreak occur in Flint? **(0.5 points)**

17.   How long did it take to switch back to Detroit water system? **(0.5 points)**

18.   When was a state of emergency issued for Flint? **(0.5 points)**

19.   What steps/laws do you think could be implemented in the future to reduce the risk of this happening again? **(0.5 points)**

20.   What should the DFW area to do prevent these same issues from happening here? **(0.5 points)**

Lab XI

Policymaking and Endangered Species

Introduction

Today, we will step out of our roles as experimental environmental scientists and step into the role of a policymaker. Most environmental problems require some amount of coordination at the policy level, both domestically and internationally. The challenge policymakers face is incorporating scientific knowledge into an intricate problem-solving process that involves social, economic, cultural, and political factors.

The international market for plant and animal trade, although vast, may not be something the average consumer considers when making purchases. Pharmaceuticals, timber products, food products, leather, and textiles are all large international markets that depend on plant and animal species for raw material. Unfortunately, there are also numerous different luxury goods that are obtained and traded illegally at the expense of endangered species—some of which may be extinct in as little as 5 years. Today we will explore the problem of rhino poaching in South Africa and the illegal trade of rhino horns on the global market.

BACKGROUND

CITES

The Convention on International Trade in Endangered Species of Wild Fauna and Flora (**CITES**) began in March 1973 as a multigovernmental agreement for the careful monitoring on international trade of wildlife and flora parts among 183 as of 2016 parties (i.e., countries that joined the agreement). The goal of CITES is to encourage international trade of animals and plants in a way that does not hinder the survival of the species in the wild. CITES prohibits the trade of a living animal or parts of an animal if the species is categorized as threatened or endangered. While CITES is not international law, each party member agrees to establish specific legislation within their country to abide by the overall goals of the agreement.[1]

Illegal Animal Trade

Despite the CITES agreement, some species are still succumbing to issues of poaching for black market trading. Many of these species have now become threatened, endangered, or critically endangered, as categorized in the International Union for Conservation of Nature (IUCN) Red List of Threatened species. The illegal animal trade has encouraged increasing numbers of poaching recently, especially among many of the *charismatic*

---

[1] UNEP. 2014. "What Is CITES?" *Convention on International Trade in Endangered Species of Wild Fauna and Flora: United Nations Environment Programme.* http://www.cites.org/eng/disc/what.php (date accessed January 13, 2014).

megafauna—large animals with widespread popular appeal that many environmental activists focus on for achieving environmental action—like Siberian tigers, rhinoceroses, or elephants. Many of the issues surrounding the illegal trade industry for animal parts are quite convoluted. For the purposes of our in-class demonstration, we will be focusing on the situation surrounding the African rhinoceros.

Rhino Poaching

There are two primary African rhino species that have been affected by poaching and illegal trade. The white rhinoceros (*Ceratotherium simum*) is listed in the IUCN Red List as near threatened,[2] while the black rhinoceros (*Diceros bicornis*) is listed as critically endangered.[3] Approximately 75% of all rhinos reside in the country of South Africa, which encompasses 93% of all white rhinos and 40% of black rhinos. Currently, the largest of the conservation issues concerns the black rhino, which is listed as vulnerable in Namibia and critically endangered in South Africa. White rhino populations have recovered from their previously critically endangered status in the last 50 years and are only considered nearly threatened today.[4]

Rhino poaching in Africa is typically associated with organized crime syndicates that recruit poachers from local impoverished communities. Locals are drawn to the significant pay out (up to $100,000 per horn) associated with poaching work. The rate of rhino poaching has increased exponentially in the last several years. Since in the last five years alone, South Africa has lost almost 5,500 rhinos to poaching. If these rates do not dramatically decline, the black rhino is expected to go extinct by 2026.[5]

The demand for poached rhino horn is largely from countries in Southeast Asia. For example, some Chinese cultures have historically believed rhino horn to possess medicinal benefits. Today, "cleansing" drinks made from rhino horn powder is one of the most prevalent of uses among the elite within society. More recently, consumers in Vietnam have begun to purchase rhino horn powder as a means to demonstrate wealth or to seek its purported medicinal qualities.[6] A rhino's horn is largely made of keratin—the same molecular component that makes up your fingernails and hair—and in reality has no legitimate medicinal uses. Thus, the demand for rhino horn in Southeast Asia is largely a result of widespread misinformation.

Like most environmental problems, the issue of rhino poaching in South Africa spans a multitude of cultural, political, and economic spheres. How can policymakers at an international level attempt to address such a complicated issue with so many different stakeholders?

---

[2] Emlise, R. 2012. "*Ceratotherium simum*." In *IUCN 2013*. IUCN Red List of Threatened Species Version 2013.2 (date accessed January 13, 2014).

[3] Emlise, R. 2012. "*Diceros bicornis*." In *IUCN 2013*. IUCN Red List of Threatened Species Version 2013.2 (date accessed January 13, 2014).

[4] Curnow, Robyn, and Teo Kermeliotis. 2013. "Could South Africa's Horn Sale End Rhino Slaughter?" *Cable News Network*. http://edition.cnn.com/2013/07/22/world/africa/south-africa-horn-sale-rhino-slaughter/index.html (date accessed January 13, 2014).

[5] Save the Rhino International. 2011. "Supply and demand: The Illegal Rhino Horn Trade." *Save the Rhino International*. http://www.savetherhino.org/rhino_info/threats_to_rhino/poaching_for_traditional_chinese_medicine/supply_and_demand_the_illegal (date accessed April 9, 2014).

[6] South Africa Department of Environmental Affairs. 2013. "South African Government Committed to Employing Various Methods to Thwart the Ongoing Poaching." *Republic of South Africa Department of Environmental Affairs website*. https://www.environment.gov.za/variousmethods_thwartpoaching (July 11).

LAB 11 WRITE-UP: THE RHINO CRISIS

In this lab, we will attempt to simulate the difficult nature of environmental policymaking by using the current rhino poaching crisis as an example.

Your instructor will assign you to be a roleplaying member of one of the following stakeholder groups:

- Government of the Republic of South Africa
- Safari clubs and Private Game Reserves
- Wildlife Conservation Non-Governmental Organizations (NGOs)
- Private Sector Conservation Groups

You will work with your group to prepare one small slideshow presentation to present your groups perspective and arguments. Before you begin, be sure to *read information about your perspective* below. Then spend some time doing Internet research to support your arguments (cite your sources!).

Your short presentation (5–10 minutes) should, at minimal, *address what you think would be the most effective solution and explain why the other solutions would not be effective.*

Some of the stakeholder groups discussed share common ideas about which conservation efforts should be taken, but may be affected by these solutions in different ways. Therefore, it is important to think about both the efficacy of your stakeholder group's proposed solutions *and* their unique perspectives of the related problems and solutions.

Specifically, consider the following when thinking of ways to organize your presentation:

1. *Frame the problem and your goal.* In other words, what do you think is the root of the poaching problem is (e.g., the Chinese black market, poverty in southern Africa, etc.) and what is your goal to solve this problem (e.g., crippling demand on the black market, allowing villagers to protect wildlife in their own habitat, etc.)?

2. Provide an *outline of your plan* to achieve your goal.

3. What are the main *problems with the other solutions* (discussed below in "Perspectives")?

4. How do the rhino poaching crisis and the potential solutions discussed affect your stakeholder group specifically?

PERSPECTIVES

The Government of the Republic of South Africa

In July 2013, the South African government submitted to CITES a proposal to lift the global ban on horn trade in an effort to cripple the rhino horn black market.[7] The idea was to reduce or eliminate the need for an illegal market with simple supply-and-demand economics: flood the market with the product in demand and the price of the product will plummet, putting the organized crime syndicates responsible for poaching out of business. The goal of the ban lift is not to encourage legal poaching, but rather to allow the South African government to sell their billion-dollar rhino horn stockpile, which they have acquired over past decades after confiscating the contraband from poachers.

The government asserts that this approach is a last resort after many expensive, failed military campaigns to target poachers. It is important to note that this plan does not involve making the unpermitted rhino hunting legal—just the trade of their horns. The central criticism of the plan argues that flooding the market with a

---

[7] Baldus, Rolf D. 2013. "South Africa May Trade Rhino Horn Legally," *Africa Indaba*, June, 11:3. http://www.africanindaba.com/2013/06/south-africa-may-trade-rhino-horn-legally-june-2013-volume-11-3/

legal horn supply may encourage poaching more because horn harvesting will appear to be an activity condoned by the government. Furthermore, because the trade of horns is currently illegal, it is difficult to estimate both the current demand and the potential demand should the ban be lifted.

Key concepts your team should research:

- Rhino horn on the black market

- Failures/successes of antipoaching security

Safari Clubs and Private Game Reserves

Governments around the world have permitted the regulated hunting of game animals for more than a century, with different degrees of success for conservation goals. In some cases, permits to kill a set number of individuals are issued to hunters annually with the intention of controlling the population (as with deer hunting). In other cases, a small number of very expensive permits are issued for "trophy hunting." Private game reserve owners and hunting groups like the Dallas Safari Club have been advocates of using a highly restricted permitting system for the trophy hunting of black rhinos as a means to raise money for conservation and possibly solve some problems within the breeding rhino population.

This method involves auctioning off a very small number of permits to wealthy hunters. Wildlife professionals handpick specific rhinos to be hunted, particularly the older "surplus males" in the population. Ideally this ensures breeding in the rhino population is not impacted, since these males are too old to breed. The removal of surplus males is meant to decrease deaths from fighting among adult males and encourage females to breed with other males in the population, increasing genetic diversity in offspring generations.

Sustainable hunting of Southern white rhino populations has been successful in the past; the number of individuals rose from 1,800 in 1968 (when permitted hunting was made legal) to 11,100 in 2005.[10] Proponents of limiting hunting argue that black rhino hunting could be done sustainably while raising millions of dollars every year for conservation efforts, which are chronically underfunded. However, some conservationists have speculated that allowing the hunting of endangered species may lead to an increase in poaching. CITES currently approves the hunting of 10 surplus male black rhinos per year.

Key concepts your team should research:

- The "surplus male problem"

- White rhino hunting successes

- The January 2014 Dallas Safari Club black rhino hunting permit auction

Wildlife Conservation NGOs

Many NGOs dedicated to the preservation of endangered species are active advocates for solving the poaching problem as an international community. The primary approaches advocated by these NGOs differ, but in general they focus on increased *habitat conservation efforts*, increased *antipoacher security*, and other methods that prevent poaching or *incapacitate the crime syndicates* responsible for selling horns on the black market. However, these NGOs generally stand firm against any antipoaching solution that involves the hunting of rhinos or sale of horns.

---

[10] Emslie, Richard. 2005. "The Debate About Rhino Hunting." *Save the Rhino.org*. http://www.savetherhino.org/rhino_info/thorny_issues/trophy_hunting/the_debate_about_rhino_hunting (date accessed January 10, 2014).

[11] Lavandera, Ed. 2014. "Debate over Dallas Safari Club Auctioning Black Rhino Hunting Permit." *CNN.com*. January 11. http://www.cnn.com/2014/01/10/us/black-rhino-hunting-permit/.

[12] NGO Pulse. 2012. "Wilderness Foundation Addresses Parliament on Rhino Poaching." *NGO Pulse website*. January 31. http://www.ngopulse.org/category/tags/rhino-poaching (date accessed January 11, 2014).

These NGOs advocate against trophy hunting to fundraise for antipoaching efforts, because they believe that approach perpetuates a message that the critically endangered species are worth more as dead trophies than as a part of their ecosystems.[11] Some NGOs, such as the International Fund for Animal Welfare, have also argued that the South African government's proposal to lift the international ban on horn trade would produce yet another communication problem in which the monetary value of rhino horns is placed over the ecological value of the species in the mind of the international community.[12]

One such NGO in this stakeholder group is the World Wildlife Fund (WWF). According to the WWF Web site, fundraising efforts for rhino conservation go toward:

> "Expanding existing *protected areas* and improving their management; establishing new protected areas; improving *security monitoring* to protect rhinos from poaching; improving local and international *law enforcement* to stop the flow of rhino horn and other illegal wildlife trade items from Africa to other regions of the world; promoting well managed *wildlife-based tourism* experiences that will also provide additional funding for conservation efforts."

> —World Wildlife Fund Web site[13]

Other NGOs advocate lessening the demand for rhino horns by educating buyers (primarily in Southeast Asia) on the environmental cost of their purchases and the lack of actual medicinal properties, whereas others encourage solutions that would prevent crime syndicates from recruiting local villagers by targeting poverty in the area.

Key concepts your team should research:

- Antipoaching security, including military force and drones

- Rhinos and eco-tourism

- Key NGOs: World Wildlife Fund, Wilderness Foundation, International Fund for Animal Welfare

Private Sector Conservation Groups

Another solution already in practice is to use large, adjacent, fenced in areas that encompass two or more private properties to protect the rhino populations. This form of conservation efforts deploys at least one armed guard for each rhino on the property to track during the day, in order to protect against unwanted poachers.

Initially, starting up a conservation reserve is costly and requires funding from outside sources, such as the WWF. As the population on site starts to increase and reach a more sustainable population size, ranchers capitalize on both the consumptive and nonconsumptive aspects of the rhino populations. Consumptive aspects include auctioning off permits for trophy hunting, while nonconsumptive gains funds from bird-watching and photography hobbyists.

Research ideas your team might want to look into:

- Rhinos and eco-tourism

- Antipoaching security

- Savé Valley Conservancy and Bubiana Conservancy in Zimbabwe

---

[13] World Wildlife Fund. "WWF Rhino Programme." WWF website. http://wwf.panda.org/what_we_do/endangered_species/rhinoceros/african_rhinos/the_african_rhino_programme/ (date accessed January 10, 2014).